GEOTEXTILE TESTING AND THE DESIGN ENGINEER

A symposium
sponsored by
ASTM Committee D-35 on
Geotextiles, Geomembranes,
and Related Products
Los Angeles, CA, 26 June 1985

ASTM SPECIAL TECHNICAL PUBLICATION 952
Joseph E. Fluet, Jr., GeoServices Inc. Consulting Engineers, editor

ASTM Publication Code Number (PCN)
04-952000-38

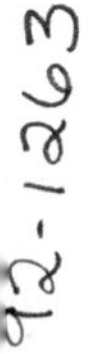

1916 Race Street, Philadelphia, PA 19103

Library of Congress Cataloging-in-Publication Data

Geotextile testing and the design engineer.

(ASTM special technical publication; 952)
Proceedings of the ASTM Symposium on Geotextile Testing and the Design Engineer.
Includes bibliographies and indexes.
"ASTM publication code number (PCN) 04-952000-38."
1. Geotextiles—Testing—Congresses. I. Fluet, Joseph E., Jr. II. ASTM Committee D-35 on Geotextiles, Geomembranes, and Related Products. III. ASTM Symposium on Geotextile Testing and the Design Engineer (1985: Los Angeles, CA) IV. Series.
TA455.G44G45 1987 620.1'97 87-12622
ISBN 0-8031-0952-0

Library of Congress Catalog Card Number: 87-12622

NOTE

The Society is not responsible, as a body, for the statements and opinions advanced in this publication

Printed in Ann Arbor, MI
July 1987

Foreword

This publication, *Geotextile Testing and the Design Engineer,* contains papers presented at the international symposium of the same name held in Los Angeles, California on 26 June 1985. The symposium was sponsored by ASTM Committee D-35 on Geotextiles, Geomembranes, and Related Products. Joseph E. Fluet, Jr., of GeoServices Inc. Consulting Engineers, presided as symposium chairman and was editor of this publication.

A Note of Appreciation to Reviewers

The quality of the papers that appear in this publication reflects not only the obvious efforts of the authors but also the unheralded, though essential, work of the reviewers. On behalf of ASTM we acknowledge with appreciation their dedication to high professional standards and their sacrifice of time and effort.

ASTM Committee on Publications

ASTM Editorial Staff

Contents

Overview

As chairman of the ASTM Symposium on Geotextile Testing and the Design Engineer, I also have the honor of being the editor of this resulting special technical publication (STP). Geotextiles have been defined by ASTM Committee D35 on Geotextiles, Geomembranes, and Related Products as "permeable textiles used with geotechnical materials as an integral part of a man-made project, structure or system." Geotextiles are part of a larger family of materials called geosynthetics, which are used by civil, geotechnical, environmental, and structural engineers in their designs. These materials include geotextiles, geomembranes, geogrids, geonets, geomats, geocomposites, and a host of other geo- terms, the list of which is growing every year as new applications are discovered.

The authors of the papers presented in this STP represent all facets of the geosynthetic discipline: academia, manufacturers, end-users, and design consultants. Furthermore, they bring with them more than a century of combined experience in a discipline which is only 10 to 15 years old.

It is important to note the difference between the geosynthetic industry, geosynthetic materials, and the geosynthetic discipline. The geosynthetic industry encompasses all the people, facilities, equipment, materials, knowledge, applications, and economics involved with geosynthetics. Geosynthetic materials, better known simply as "geosynthetics," are the products developed by researchers, made by manufacturers, sold by salesmen, and purchased by end-users to be installed by contractors in applications designed by engineers. The geosynthetic discipline is that body of knowledge which has been developed to make it possible to design with geosynthetics using rational engineering design methods and to explain the success of these materials using modern scientific concepts and principles. This discipline therefore refers to those scientists and engineers who are concerned with the development of geosynthetic theory and understanding as well as the design of these applications.

This STP is directed at the heart of the geosynthetic discipline. It describes the tests used by scientists and engineers to develop and then specify geosynthetics, and it then discusses the design of applications where those test results are used. One consistent goal of this discipline has been to develop design methods and approaches which are consistent with, and, whenever possible, extensions of existing engineering design methods. As a consequence, many of today's geosynthetic design approaches will seem very familiar to civil, geotechnical, environmental, and structural engineering designers. This has been no mean feat since this discipline, as reflected in the published literature, began only recently, (although it describes an industry which is almost as old as civilization itself: the Babylonians constructed soil-reinforced "ziggurats" more than 3000 years ago, and the Chinese have used reeds to construct a type of corduroy road since prehistoric times). Only recently have we learned to understand the engineering functions of geosynthetics, to calculate the engineering forces involved, and therefore to design with and predict the success of these materials in various engineering applications. It has been this understanding, that is, the engineering discipline, which has prompted the acceptance of geosynthetics as engineering "state-of-practice" materials. As a result of this growing acceptance, combined with manufacturer marketing efforts, more than 200 million square metres of geosynthetics are now being sold per year in North America, and the market is growing at roughly a 15 to 20% annual rate of increase.

This rapid growth has been plagued with many of the problems which beset any emerging discipline, and one problem which has been particularly troublesome is interdisciplinary communications. Of necessity, the geosynthetic discipline has representation from civil, geotechnical, environmental, structural, industrial, mechanical, chemical, textile, and plastics engineering, as well as the scientific fields of chemistry, biology, physics, and others. Unfortunately, each of these disciplines has its own jargon, and their language and concepts are sometimes confusing to, if not in conflict with, each other. For example, the term "filter," as used by mechanical engineers, chemists, and chemical engineers, refers to a completely different concept than that used by civil and geotechnical engineers—the former expect filters to clog and to be periodically replaced, while the latter devote considerable energy to designing filters that will not clog. Similarly, textile engineers struggle with designing textiles for the clothing industry which are smooth to the touch and stretch easily with body movement, while geotechnical engineers yearn for geotextiles with high friction angles and high modulus, which means they must be rough and unyielding.

The only answer to this type of apparent impasse is to increase the avenues and opportunities for communication between the parties involved.

This STP is one such opportunity. It discusses both the textile industry tests and the relevance of those tests to the design engineer. The tests are grouped by the broad types of properties which they measure—mechanical, hydraulic, and endurance—and each category is followed by papers which describe the relevance of the test results to the design engineer. Additionally, there is a section on the future of geosynthetic tests and applications, which shows clearly that, although we have come very far, we still have a long way to go.

One other highlight of the STP is the ASTM position paper prepared by Christopher, Carroll, and Suits. This paper is intended as an interim guide for those laboratories which cannot wait for the publication of ASTM standards which are currently under development. ASTM Committee D35 (and its predecessors, ASTM Committees D13.61 and D18.19) has already published several standards, and many others are currently in various stages of development. This position paper describes the current (1985) state of development of those standards which are not yet complete. No suggestion or representation is made concerning the relevancy, accuracy, appropriateness, or completeness of the tests described in the position paper. In fact, the tests described may bear little resemblance to the standards that are eventually published. Nonetheless, the tests described are, in the opinion of this editor as well as the authors, representative of the tests most commonly in use at this time.

No introduction to an STP which results from the work completed by ASTM Committee D35 would be complete without a word about the members of this committee. Many of these scientists and engineers have faithfully participated in D35 (and its predecessor committees) since 1978, and have struggled with every painstaking step in the process of developing new standards. As of January 1986, the following geosynthetic standards have been passed by ASTM:

Standard Number	Title
D 4354-84	Practice for Sampling of Geotextiles for Testing
D 4533-85	Test Method for Trapezoid Tearing Strength of Geotextiles
D 4595-86	Test Method for Tensile Properties of Geotextiles by the Wide Width Strip Method
D 4632-86	Test Method for Breaking Load and Elongation of Geotextiles (Grab Method)
D 4355-84	Test Method for Deterioration of Geotextiles From Exposure to Ultraviolet Light and Water (Xenon-Arc Type Apparatus)
D 4594-86	Test Method for Effects of Temperature on Stability of Geotextiles
D 4491-85	Test Method for Water Permeability of Geotextiles by Permittivity

D 4437-84	Practice for Determining the Integrity of Field Seams Used in Joining Flexible Polymeric Sheet Geomembranes
D 4545-86	Practice for Determining the Integrity of Factory Seams Used in Joining Manufactured Flexible Sheet Geomembranes
D 4439-85	Terminology for Geotextiles

To those veterans of the geosynthetic standards wars, as well as to those newer members who are helping to continue the campaign, I offer heartfelt thanks.

Joseph E. Fluet, Jr.

GeoServices Inc. Consulting Engineers, Boynton Beach, FL 33435; symposium chairman and editor

Drainage and Erosion Control

Robert G. Carroll, Jr.[1]

Hydraulic Properties of Geotextiles

REFERENCE: Carroll, R. G., Jr., **"Hydraulic Properties of Geotextiles,"** *Geotextile Testing and the Design Engineer, ASTM STP 952,* J. E. Fluet, Jr., Ed., American Society for Testing and Materials, Philadelphia, 1987, pp. 7-20.

ABSTRACT: Hydraulic properties of geotextiles are controlled by the geotextiles' pore sizes, pore size distribution, and porosity. Because porometry characteristics of fabric are difficult to measure precisely, index tests have been developed that relate other fabric characteristics to hydraulic performance. Apparent opening size, permittivity, gradient ratio, and transmissivity are hydraulic properties currently being used as performance criteria for geotextiles. ASTM Committee D-35 on Geotextiles, Geomembranes, and Related Products is developing standard procedures for characterizing these properties.

KEY WORDS: geotextiles, pore size, apparent opening size, equivalent opening size, permeability, permittivity, gradient ratio, transmissivity, test methods, ASTM

Hydraulic properties of geotextiles are those that relate directly to filtration and drainage functions of geotextiles. Hydraulic properties fall into two general categories:

1. *Filtration*—pertaining to the filtration or separation function of a geotextile when used to replace graded aggregate filters in subsurface drains.
2. *Drainage*—pertaining to a geotextile's ability to transport water within its plane when used as the drain replacing aggregate or sand as the water conductor.

The filtration function of a geotextile requires that it have pore size sufficiently small to retain erodible soil particles and permeability adequate to allow the free escape of seepage from the protected soil. These filtration properties are controlled by the pore sizes, pore size distribution, and porosity of the fabric. Characterization of geotextile porometry is a complex and difficult task because of the size and magnitude of pores in a fabric. As a result, a simple index test has been devised that measures geotextile porometry in terms of particle retention ability. As the name implies, "apparent opening size" indicates the apparent maximum pore diameter within a geotextile.

The permeability of geotextiles is determined by measuring (under standard conditions) the flow rate of water through the fabric in a direction normal to the plane of the fabric. This flow rate is used to calculate two permeability factors which are routinely used with respect to geotextiles.

1. *Darcy's permeability coefficient*—compares with Darcy's permeability coefficient of protected soil to determine compatibility.
2. *Permittivity*—allows for comparison of flow capability between fabrics.

Apparent opening size (AOS) and permeability fulfill the conventional criteria for selecting a filter media. However, these properties do not relate directly to the clogging potential of a geotextile. Clogging behavior must be evaluated in a soil-fabric system that simulates use condi-

[1]Marketing manager, Tensar Corp., Morrow, GA 30260.

tions. A Gradient Ratio Test is one such method where a soil-fabric permeameter system is used to monitor soil-fabric interaction. Results from this test are interpreted in terms of clogging performance.

When the geotextile itself is used as a drain, that is, a conduit to transport water (or air) within its plane, the in-plane flow capacity of the fabric must be measured. The hydraulic transmissivity method is a fabric permeability test developed for this purpose. Transmissivity results are a direct measure of flow rate per unit thickness of a geotextile under specified conditions of stress and hydraulic gradient.

This paper provides a review of these key hydraulic properties for geotextiles and addresses the following issues with respect to each:

1. Need or importance.
2. Method of measure or evaluation (ASTM Committee D-35 standards).
3. Significance of use.
4. Deficiencies in state of practice and development needs.

Filtration Properties

Filtration Criteria

Regardless of the filter medium chosen for drainage applications, it must meet two conflicting requirements to assure optimum performance:

1. *Retention*—the filter must have a pore structure fine enough to retain erodable soils.
2. *Permeability*—The filter must maintain adequate permeability so that seepage can escape freely from the protected soil; clogging resistance is inherent to this requirement.

Grain-size distribution of a graded aggregate filter creates its pore structure which, in turn, controls filtration performance. There are universally accepted criteria for specifying the grain-size distribution of aggregate filters that relate particle size of a graded aggregate to that of the protected soil [*1*]. These criteria, based on theoretical relationships between particle size, pore size, and retention ability of granular materials, have proven adequate through decades of use.

There are no well-established filter criteria for geotextiles. Filtration performance of a geotextile is controlled by its fiber structure. Fiber structure determines pore sizes, pore distribution, and porosity, which in turn control fabric retention ability and permeability. The ideal retention criteria for fabrics should specify the appropriate pore structure to eliminate piping through the fabric, to provide adequate fabric seepage rate, and to assure clogging resistance. But, an accurate measure of pore structure in porous media is difficult to obtain. Though numerous tests have been developed, no method has been universally accepted. The next best alternative to an accurate measure of filter pores is an index test (or tests) that relates pore characteristics to filtration performance. Such index values are the basis for filter media selection in most filtration applications using analytically and/or empirically based criteria.

Retention Ability and Equivalent Opening Size

In the late 1960s, Calhoun [*2*] performed research on "filter cloths" at the Corps of Engineers (COE) Waterways Experiment Station. The objective of that COE project was to develop acceptance specifications and design criteria for plastic filter cloths used in filtration-drainage applications. Calhoun evaluated several fabrics, most of which were woven monofilament, with one woven multifilament and one nonwoven fabric also included. At the time the study began, woven monofilament fabrics were the only type used in the United States for filtration-drainage applications. These woven monofilament fabrics resembled screen mesh, though their yarn spacing varied somewhat and pore openings were not square. The woven multifilament and nonwoven fabrics were unlike screen mesh; they had no discrete openings and their pore structures were apparently very fine.

Calhoun developed a test for equivalent opening size (EOS) to characterize the soil particle retention ability of the various fabrics. The test involved sieving rounded sand particles of a specified size through the fabric to determine that fraction of particle sizes for which 5% or less, by weight, passed through the cloth. The EOS was defined as the "retained on" size of that fraction expressed as a U.S. Standard Sieve Number (for example, No. 70 or 0.210 mm). Assuming that fabrics and screen mesh have comparable retention ability, the EOS was a rational means of correlating fabric pore structure to an equivalent screen mesh size. EOS could then be used to indicate a fabric's retention ability.

EOS Versus Pore Structure

It is imperative to note that EOS values do not accurately define fabric pore sizes, pore structure, or filtration ability. For decades, filter media producers and users have adopted various techniques similar to the EOS test for measuring the retention or filtration efficiency of their products. Shoemaker [*3*] notes that most filter manufacturers have adopted "micron rating" techniques, but the method for arriving at the "rating" varies with the manufacturer and the product. The concept of rating is very helpful when effecting a relative ranking of retention characteristics of similar products for one manufacturer. A cartridge that has a rating of 5 μm is presumably more retentive than one with a 50-μm rating. In comparing like products, such as cartridges from two different manufacturers, the numbers may not be equivalent. When comparing unlike products, such as cartridges versus felt versus paper, it is difficult to justify absolute numbers.

Visual examination of different fabrics with the same EOS illustrates the variety of pore structure and porosity that can exist despite common EOS values. Figures 1A and 1B show a woven monofilament and a nonwoven fabric, respectively, both with EOS = 0.210 to 0.177 mm. Figures 2A and 2B show a woven monofilament and a woven slit film fabric, respectively, both with EOS = 0.595 to 0.420 mm. Note the obvious differences in pore structure despite comparable EOS values.

The EOS test provides a method for determining relative size of the maximum "straight through" openings in a fabric. And as such EOS values provide an indication of the minimum particle size the fabric can retain. EOS does not, however, indicate absolute retention efficiency for geotextiles; that is, a woven monofilament and nonwoven fabric, both with EOS = 0.210 mm, will not have the same pore structure and therefore will not provide the same filtration efficiency for all particle sizes [*4*]. The difference in filtration efficiency does not discount the validity of a retention criterion using EOS. It merely indicates that certain fabric pore structures tend to exhibit greater retention efficiencies than others (for example, nonwoven versus woven, respectively).

AOS Versus EOS

Calhoun's EOS test gained wide usage in the geotextile industry for product characterization and specification. Through a decade of use, the original test method has undergone refinements to the apparatus and revisions of the test results. Today, there are three documented methods for characterizing geotextile opening sizes: (1) the procedure defined by Calhoun [*2*] using graded sand particles; (2) the modified version defined in the COE Civil Works Guide Specification for Plastic Filter Fabric [*5*] which uses graded glass beads; and (3) a revised version of the previous two being prepared for standardization by ASTM.

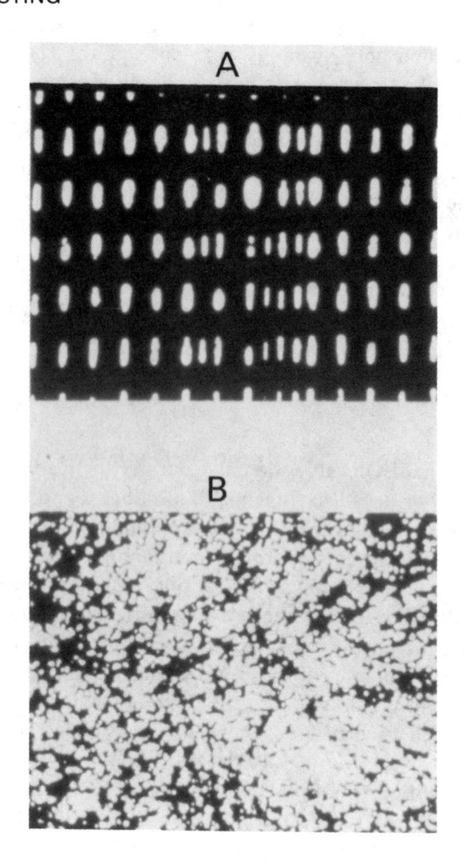

FIG. 1—(A) *Woven monofilament fabric with EOS = No. 70 to No. 80.* (B) *Nonwoven fabric with EOS = No. 70 to No. 80.*

The ASTM method measures apparent opening size (AOS), so named because the results of this test do not qualify equivalence between a fabric and a U.S. Standard Sieve or another fabric. Instead the test indicates the "apparent" largest pore sizes (straight through pores) of the fabric. In this test, a geotextile specimen is placed in a sieve frame, sized glass beads are placed on the geotextile surface, the particles are placed on the geotextile surface, and the geotextile and frame are shaken laterally so that the jarring motion will entice the beads or particles to pass through the specimen. The procedure is repeated on the same geotextile specimen with various size glass beads or particles until its equivalent or apparent opening size has been determined. Note that equivalent or apparent opening size is that bead or particle size (minimum diameter of bead or particle size range) for which 5% or less of the beads pass through the fabric.

Testing laboratories have noted that the sieving process with glass beads typically yields lower EOS values than sieving with sand particles, that is, $EOS_{glass} = 0.297$ mm and $EOS_{sand} = 0.210$ mm for the same fabric. Variability between these tests is attributed to differences between glass beads and sand particles, for example, particle roundness, static potential, etc.[2]

[2]Personal communication with B. R. Christopher of Soil Testing Services, Northbrook, IL, Jan. 1982.

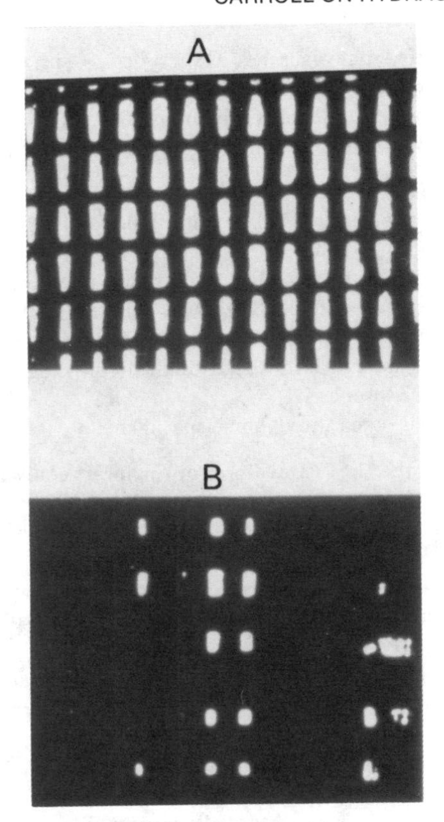

FIG. 2—(A) *Woven monofilament fabric with EOS = No. 30 to No. 40.* (B) *Woven slit film fabric with EOS = No. 30 to No. 40.*

The ASTM group initiated a round-robin program to evaluate the AOS testing concept and its reproducibility. The program addressed the following factors which were identified as problem areas with respect to reproducibility:

1. Surface film on the geotextile.
2. Static buildup on glass beads.
3. Order of bead range sieving (fine to coarse or coarse to fine).
4. Shaking time.
5. Variability between specimens.

In some instances, the manufacturing process leaves a surface coating on the geotextile which may act to clog some of the openings. This leads to erroneous test results in that beads that would normally pass through the geotextile cannot. However, for most geotextiles which have this coating, it has been found that soaking the fabric in water will remove this coating, allowing the appropriate size beads to pass.

The buildup of static electricity causes glass beads to cling to the geotextile rather than pass through, again leading to erroneous results. "Static masters" added to sieve frame walls will reduce or eliminate the buildup of static electricity. Other static eliminators in the form of aerosol sprays applied on the fabric have also been found effective in eliminating the static electricity problem.

Round-robin testing revealed that when the AOS test begins with coarse glass beads, certain fabric pores that would pass finer beads might become clogged. The round-robin program also noted no significant difference in test results for varying sieve times from 5 to 20 min.

Taking all these factors into consideration, the proposed method for AOS will require the following:

1. A 1-h presoak of the specimen in distilled water followed by drying at 30 ± 5°C until a constant weight is reached.
2. Drying of the glass beads at the same temperature will also be required.
3. Static eliminator devices (for example, Static Master) are to be used in the sieving frame to decrease or eliminate the buildup of static electricity on the beads.
4. The sieving process will start with fine beads first, progressing to the coarse beads.
5. A 10-min shaking time is proposed.
6. All size glass beads are shaken through one specimen.

This proposal differs from the U.S. Army Corps of Engineers equivalent opening size procedure in the use of a presoak period, Static Masters, and shorter sieving time. The ASTM Standard Test Method for Determining the Apparent Opening Size of a Geotextile is currently being balloted in ASTM.

Several other techniques have been proposed for evaluating fabric pore structure, including the use of image analyzers and mercury intrusion. In general, these techniques provide a reasonable indication of fabric pore sizes and/or structure, but they require complex procedures and expensive equipment that limit their utility to research rather than routine use by agencies and consultants.

Permeability

Permeability of a geotextile must be substantially greater than that of the protected soil so that water can pass freely from the soil through the fabric without buildup of hydrostatic pressure. High fabric permeability also infers that partial clogging will not reduce fabric permeability to a critical level, that is, below that of the protected soil. Permeability values for fabric and soil are necessary to assess compatibility between the two. Soil permeability is typically determined by measuring the flow rate of water through a soil sample of specified height and cross-sectional area. Soil permeability is then expressed in terms of a coefficient calculated using Darcy's law regarding laminar flow through porous media

$$q = kiA = k \frac{\Delta h}{L_f} A \tag{1}$$

where

q = flow per unit time,
k = Darcy's coefficient of permeability,
i = hydraulic gradient,
A = total cross-sectional area available for flow,
Δh = change in hydraulic head or hydraulic head loss through the fabric, and
L_f = length of flow path (fabric thickness) over which Δh occurs.

The geotextile industry has applied the same principles of flow through porous media to characterize the permeability of fabrics. Using test apparatus modeled after soil permeameters with both falling constant head testing techniques, a fabric's permeability coefficient can be determined according to Darcy's law. These fabric permeability coefficients are then compared to that of the soil to assure compatibility.

Researchers have disputed the validity of computing a Darcy coefficient for fabrics from such permeameter testing. The fabrics are very thin relative to soil thickness used in conventional soil permeameters, and flow through the fabric could be turbulent even with low head pressures. Therefore, Darcy's theory may not apply to these test conditions. Despite these inconsistencies between test conditions and theory, an "apparent" Darcy coefficient can be determined for fabrics. If turbulent flow conditions are present during testing, then the "k"-value measured for a fabric will be conservative, that is, lower than the actual k. An apparent Darcy coefficient can also be determined for fabrics using results from ASTM Test Method for Air Permeability of Textile Fabrics [D 737-75 (1980)] [*6*].

The problem with a Darcy permeability coefficient for fabrics, from a practical point of view, is the value's dependency on fabric thickness. A review of Darcy's law shows the Darcy coefficient of permeability t to be directly proportional to the length of flow path, that is, fabric thickness. Therefore, it is possible for a thin geotextile (0.5 mm) and a thick geotextile [2.5 mm (0.100 in.)] to have comparable permeability coefficients with dramatically different flow rates through each fabric. As a result, Darcy permeability coefficients do not provide a fair comparison of flow capabilities between fabrics. Such comparisons must be made using a permeability factor that is independent of fabric thickness.

In response to this need for a thickness independent permeability coefficient, the ASTM Committee on Geotextiles, Geomembranes, and Related Products has developed a test for fabric permeability. The Committee has defined permittivity of a geotextile as the volumetric flow rate of water per unit cross-sectional area per unit head under laminar flow conditions, in the normal direction through a geotextile. The units are seconds to the minus one ($cm^3/s \cdot cm^3 = s^{-1}$). Calculated using the Darcy equation, permittivity is actually the coefficient of permeability divided by fabric thickness (k/L). Permittivity values provide a fair comparison of permeability between fabrics and also allow for calculation of a Darcy's coefficient of permeability if necessary.

A variety of permeability test apparatus and procedures were reviewed by ASTM. A round-robin program, involving eleven laboratories, was conducted to evaluate the effect of test parameters on variability of results. Results of that round-robin provided the following conclusions:

1. All permeameter apparatus used (for example, constant and falling head) provided reasonable results as long as the apparatus did not control flow through the system.
2. Constant head permeameters should be run with a water head of 50 mm above the fabric to approach laminar flow conditions.
3. Falling head permeameter tests should be run with an equivalent head drop to the constant head test to provide comparable results between the two, that is, Δh from 80 to 20 mm.
4. Deaired water (6 ppm dissolved oxygen max) should be used to assure reproducibility of results.
5. All permeability results should be normalized to 20°C.

The ASTM Standard Test Method for Water Permeability of Geotextiles by Permittivity is currently published under the designation of D 4491-85.

Many other fabric properties have been proposed as indicators of fabric permeability (for example, porosity and percent open area). Although these properties do have some influence on fabric permeability, no firm correlations have ever been established. And no reliable correlations have ever been established between those properties and fabric filtration performance [*4*].

The effect of fabric compressibility on fabric permeability is still another concern of researchers. Schober and Teindl [*7*] show that a compressive force of 100 kPa (146 psi) can reduce the *k*-value of highly compressible needle-felt fabrics by factors ranging between 2 and 8. This compressive force is roughly equal to fabric buried beneath 150 ft of dense soil. Most drains are installed to relatively shallow depths where compressive force on the fabric filter is relatively low and the potential for reduced permeability is insignificant. If ground pressures on the fabric filter are expected to be extremely high, then permeability measurements should be determined on a fabric under the appropriate compressive force. ASTM is in the early stages of development of a standard on permeability under compressive force.

Clogging Resistance and the Gradient Ratio Test

The retention and permeability requirements of a geotextile can be specified in terms of AOS and permittivity values. These criteria are fully adequate in a majority of drainage applications where the filter's clogging potential is very low or in noncritical drainage applications [*4*]. When the potential for filter clogging is significant or the drainage application should be classified as critical, the clogging resistance of filter fabrics should be evaluated to assure adequate long-term filter performance.

Calhoun [*2*] performed clogging tests to determine the degree of fabric clogging that might be experienced by fabric in contact with a gap-graded soil. The clogging test employed a permeameter device as shown in Fig. 3. Hydraulic gradient data from the soil-fabric permeameters were analyzed to determine the clogging potential of a fabric. The analysis made use of a ratio of the hydraulic gradient across the fabric plus an adjacent 1 in. of soil to the hydraulic gradient for the entire system, that is, the "clogging ratio." A clogging ratio greater than 1 signifies fabric clogging. Calhoun's results varied, depending on fabric and soil gradation, but no clogging ratios exceeded 2. The COE [*5*] later established a maximum acceptable clogging ratio of 3.0, based on these and subsequent clogging test evaluations.

Haliburton [*8*] investigated clogging resistance of woven and nonwoven fabrics using a hy-

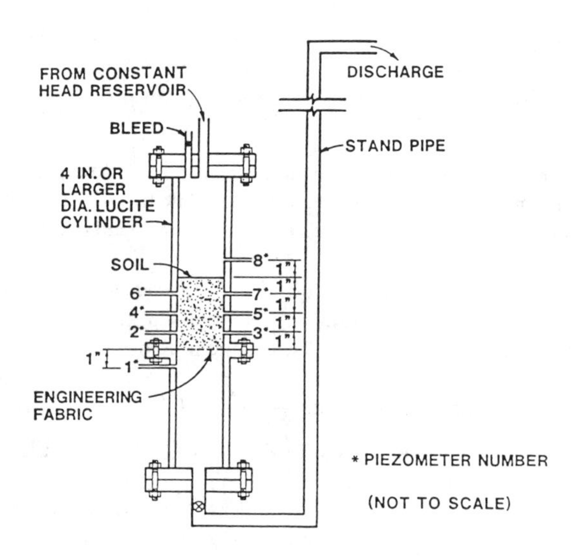

FIG. 3—*Constant-head permeameter used to measure filtration and clogging behavior in soil-fabric systems (from Calhoun* [2]*).*

draulic gradient analysis approach similar to Calhoun's. Haliburton and Wood based clogging performance on a "gradient ratio" value which is the hydraulic gradient through fabric plus the adjacent 25 mm of soil divided by the hydraulic gradient through the adjacent 50 mm (2 in.) of soil (Fig. 4). The soil used was gap-graded to provide the maximum potential of soil piping and filter clogging. In addition, the tests were run under high hydraulic gradients to cause the maximum potential for soil piping. The results of Haliburton and Wood revealed dramatic performance differences between the fabrics tested versus silt content in the gap-graded soil. Clogging potential increased for all fabrics as the silt content increased in the protected soil (see Fig. 5).

Soil-fabric clogging tests performed by the author revealed a similar performance contrast between high and low hydraulic gradient testing [*4*]. A permeameter device similar to those in previous clogging studies was used to generate gradient ratios at various system hydraulic gradients. Woven and nonwoven fabrics and a graded aggregate filter were evaluated using a well-graded silty sand (15% finer than 0.074 mm sieve) as the protected soil. Table 1 lists filter properties and the gradient ratios measured at various hydraulic gradients.

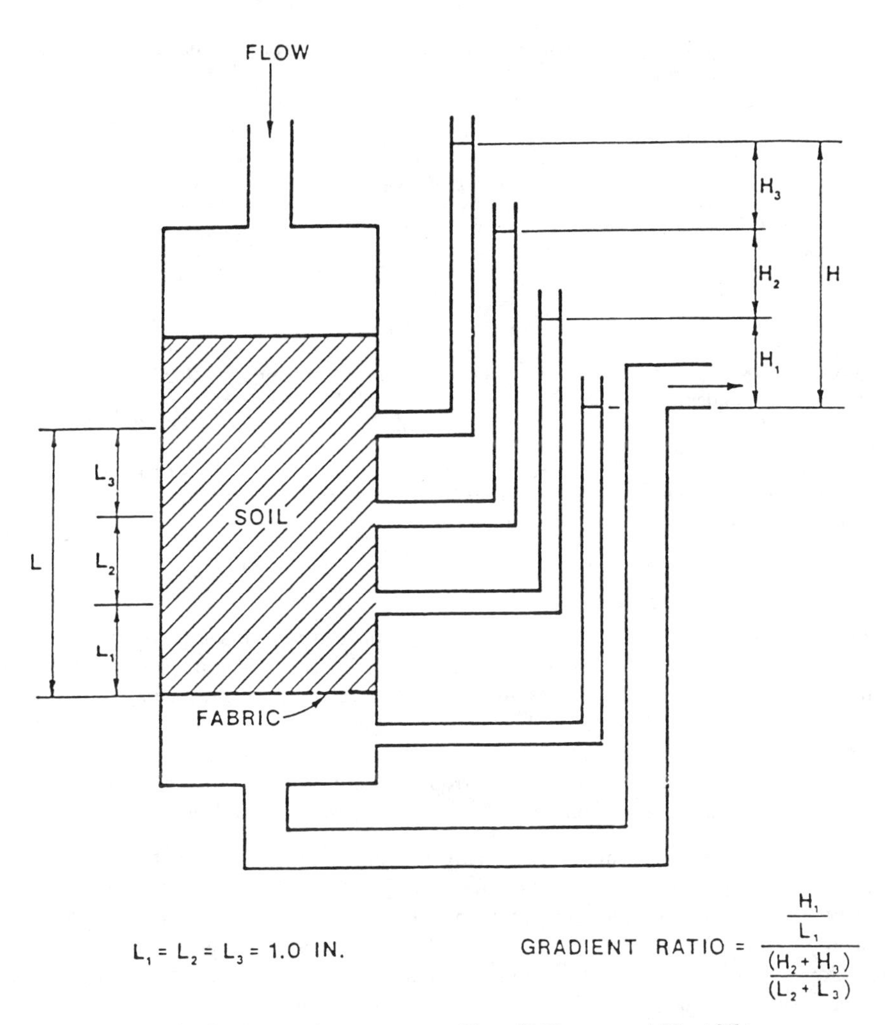

FIG. 4—*Gradient ratio permeameter (from Haliburton and Wood* [8]*).*

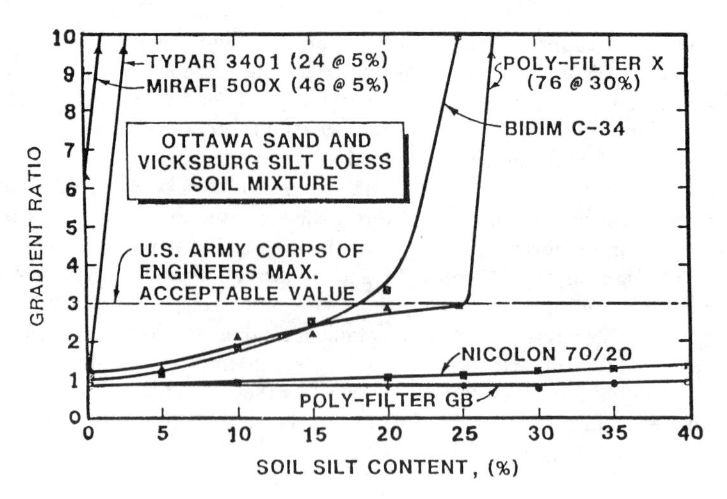

FIG. 5—*Results of gradient ratio testing for various engineering fabrics at different soil silt contents (from Haliburton and Wood* [8]*).*

TABLE 1—*Soil-fabric clogging test results.*

Filter Identification			Gradient Ratio[a]		
Type	EOS, U.S. Standard Sieve No.	k, cm/s	$i_s = 1$	$i_s = 3$	$i_s = 5$
Graded aggregate	N/A	>0.10	0.39	0.66	1.06
Nonwoven (heat-bonded)	100	0.049	0.37	0.84	1.61
Nonwoven (heat-bonded)	20	0.094	0.12	0.42	0.85
Nonwoven (needle-punched)	60	0.283	0.52	1.29	2.42
Woven (slit film)	50	0.001	0.45	1.05	2.09

[a]Gradient ratio values listed for system hydraulic gradients of 1, 3, and 5.

Note that the gradient ratios are approximately 1 or below until system hydraulic gradients become equal to or greater than 3. Fabric clogging or soil infiltration are apparently not significant when system hydraulic gradients are 3 or below. Clogging becomes more noticeable as a system hydraulic gradient increases beyond 3. Neither permeability nor EOS of the fabrics tested showed a relationship to soil-fabric system performance.

Several general conclusions can be drawn from the combined results of these clogging studies by Calhoun, Haliburton and Wood, and Carroll. These conclusions follow:

1. Fabric EOS and permeability coefficients do not indicate clogging potential.
2. All filter media are likely to experience some degree of clogging due to soil infiltration.
3. Well-graded soils are not prone to piping. High hydraulic gradients may, however, cause infiltration of well-graded soils into a filter media.
4. Gap-graded soils are prone to soil piping and subsequent filter clogging. High hydraulic gradients maximize the potential for piping in gap-graded soils.

Soil type and hydraulic gradients used in a soil-fabric permeability test have significant impact on the test results, that is, system permeability versus time and gradient ratio values. Therefore, clogging behavior of a geotextile should be evaluated in a test that simulates use conditions. For a filtration-drainage application, this would include testing a soil-fabric system in a permeameter apparatus similar to those described earlier. The soil and hydraulic gradient conditions used in testing should duplicate expected field conditions. Test parameters that deviate significantly from the use conditions will not provide a useful performance evaluation.

Koerner and Ko [9] suggest that a similar soil fabric permeability test be run and that the permeability versus time results be used to indicate system clogging. Nearly all tests of this type demonstrate a decrease in system permeability with time. Depending on the soil conditions (for example, gradation and relative density), the decrease in permeability may level off or most likely decrease at a decreasing rate indicating an apparent asymptote condition. The final permeability and length of time required to reach that apparent asymptote are said to be indicators of clogging potential.

The problem with this method of evaluation is the inability to isolate the factors controlling the change in permeability. Soil densification (consolidation), leachate from the water, microorganisms forming within the soil, and air coming out of suspension in the system can all cause a decrease in system permeability which might result in the invalid conclusion that the fabric is clogging.

The gradient ratio analysis attempts to normalize these extraneous influencing factors, providing a more accurate assessment of the fabrics' effects on system performance.

The ASTM Committee on Geotextiles, Geomembranes, and Related Products is evaluating a modified version of Calhoun's and Haliburton and Woods' Clogging or Gradient Ratio Test in a round-robin program. The procedure calls for placing the geotextile and 76 mm (3 in.) of a select soil in a 100-mm-diameter permeameter apparatus. Several piezometers or manometers are placed at various heights above the geotextile in the soil, as well as below the geotextile. A constant head of water is applied to cause flow through the system under specified hydraulic gradients. Head losses are measured at each piezometer location at regular time intervals up to 24 h. The gradient ratio values reported are calculated from measurements taken at 24 h.

Preliminary results from this round-robin on gradient ratio testing indicate reasonable reproducibility under controlled lab conditions. As with any soil test the specimen preparation and test setup are critical to the success and reproducibility of this gradient ratio test. Considerably more inner lab testing will be required to validate the reliability of this test as an ASTM standard. Despite its formative stage of ASTM development, the Gradient Ratio Test has been and continues to be used with good success within the industry to predict geotextile clogging potential.

The majority of filter criteria for geotextiles in use today employ EOS (AOS) and permeability criteria. The state of the art in this field of filter fabrics suggests that these same criteria expressed in terms of AOS and permittivity are sufficient for the majority of routine filter fabric applications. Gradient ratio values for geotextiles allow for a relative comparison between filter fabrics and selection of the one with best clogging resistance. The ASTM tests for AOS, permittivity, and gradient ratio reflect the best and most practicable methods available for evaluating these hydraulic properties of geotextiles.

Drainage Properties

Planar Flow and Transmissivity

Some geotextiles have sufficient thickness and interconnecting pore structures to allow water or gas to flow within the fabric's plane. Because of this flow capability or planar permeability, these fabrics can be used as a drain to collect and transport water or gas to an outlet or collector. As with conventional aggregate drains, a geotextile drain must have sufficient flow capacity to transport all the seepage it collects along a specified distance. Flow capacities for aggregate drains are calculated using aggregate permeability coefficients and drain cross-sectional areas. Planar flow capacity of geotextiles is quantified in a similar fashion and expressed as "transmissivity."

Unlike aggregate drains, the geotextiles that have significant transmissivity are typically lofty nonwoven fabrics that are quite compressible under stress. As these geotextiles compress, their transmissivity decreases. Therefore, it is important to quantify fabric planar flow capacity under stress comparable to that which it would experience in use. Transmissivity, then, is defined as the planar coefficient of permeability times the thickness of the geotextile at the particular stress under which it will function.

Note that creep due to long-term compression loading conditions might also result in significant reduction in transmissivity. Tests under applied stress should be performed over a sufficient time period to evaluate the influence of creep. The hydraulic gradient conditions under which the test is run will also influence results. The test should incorporate hydraulic gradients expected in field use.

Transmissivity Tests

Numerous tests methods have been developed for experimental measurement of transmissivity. The majority of these methods can be categorized as either parallel or radial flow tests [*10*]. Parallel test devices provide a longitudinal flow path so that stream lines of flow through the geotextile are generally parallel. Parallel stream lines resemble flow through a laboratory soil permeameter and similarly Darcy's law is used to calculate the transmissivity. Radial test devices allow flow to enter the geotextile at the inner circumference and exit through the outer circumference. The stream line of flow therefore radiate from the center of a circular disk of geotextile outward in all directions. Darcy's law is also used to calculate transmissivity, but integration over various radii is required in a manner similar to the "well equation" for flow to a central well point.

The ASTM Committee on Geotextiles, Geomembranes, and Related Products has reviewed the available methods for measuring "hydraulic transmissivity" and selected a parallel flow test to develop as a standard. The method provides a procedure for determining geotextile transmissivity under specified constant hydraulic head conditions and under varying compressive stresses ranging from 10 kPa (1.5 psi) to a maximum as specified by the user.

A schematic of the test apparatus is shown in Fig. 6. The loading mechanism can be either static weights, pneumatic bellows, hydraulic piston, or any other system that can suitably apply and sustain the specified uniform loads up to 240 kPa (35 psi) over the entire specimen surface. The specimen size is also critical. In order to assure adequate flow length and to reduce boundary effects, the specimens should be at least 305 mm wide by 305 mm long.

The test procedure yields flow measurements under specified conditions that can be used to calculate a value for hydraulic transmissivity as follows

$$\Theta = \frac{qL}{wh} \tag{2}$$

where

Θ = hydraulic transmissivity,
q = quantity of fluid discharge per unit time,
L = total length of specimen,

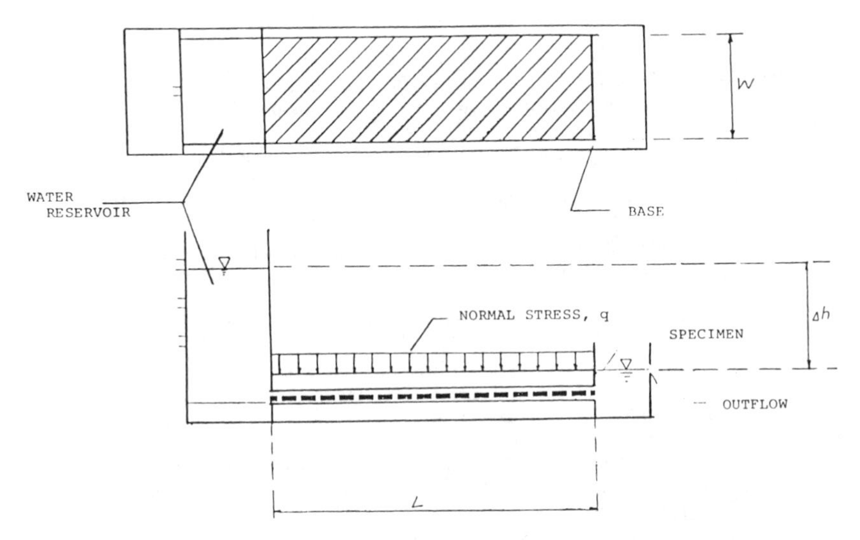

FIG. 6—*Hydraulic transmissivity testing device.*

w = total width of specimen, and
h = difference in hydraulic head across the specimen.

This test method is in the formative stages of development in ASTM, but the procedural points and apparatus described have been widely used in the industry to measure hydraulic transmissivity for both characterization and design purposes. The apparatus and method is also used to measure flow capacity and transmissivity of prefabricated drainage composites such as nets and waffle cores with fabric laminated to one or both sides.

Conclusions

This paper presents a review of important hydraulic properties of geotextiles and test methods for measuring these properties. EOS and coefficient of permeability (Darcy's) are the most widely used properties in specifications for geotextiles used in filtration applications. As with most properties and tests for soils and construction material, there is debate over the relevance of these geotextile properties to performance capabilities. And the tests for measuring these properties are also strongly debated regarding their relevance to performance and reproducibility of test results.

ASTM has recognized the concern within the industry and devoted a concerted effort to develop tests that measure the key hydraulic properties of geotextiles. Many properties and even more test methods have been reviewed for technical merit and utility. The result is the four hydraulic properties and their test methods discussed in this paper, that is, AOS, permittivity, gradient ratio, and transmissivity.

This is certainly not the end of technical development for hydraulic properties of geotextiles. Other relevant properties and test methods have been and are being discovered and developed outside of ASTM. Refinements to our current practices will only come through experience in design, specification, testing, performance monitoring, and technical exchange of this information through forums such as ASTM. As the state of the art grows and the state of the practice matures, the hydraulic properties of geotextiles will become at least as well-defined as similar properties for soil and rock.

References

[1] Karpoff, K. P., "The Use of Laboratory Tests to Develop Design Criteria for Protective Filters," *Proceedings,* American Society for Testing and Materials, Philadelphia, Vol. 55, p. 1183, 1955.
[2] Calhoun, Charles C., "Development of Design Criteria and Acceptance Specifications for Plastic Filter Cloth," Technical Report S-72-7, Army Waterways Experiment Station, Vicksburg, MI, June 1972, pp. 6-55.
[3] Shoemaker, W., "The Spectrum of Filter Media," in *Filtration and Separation,* Upland Press Ltd., Croydon, CR91LB, Jan./Feb. 1975.
[4] Carroll, R. G., "Geotextile Filter Criteria," Transportation Research Record 916, Engineering Fabrics in Transportation Construction, Transportation Research Board, Washington, DC, 1983, pp. 46-53.
[5] "Civil Works Construction Guide Specification for Plastic Filter Fabric, CW 02215," Department of the Army, Corps of Engineers, Office of Chief Engineer, Nov. 1977, pp. ii.
[6] Carroll, R. G., "Determination of Permeability Coefficients for Geotextiles," *Geotechnical Testing Journal,* American Society of Testing Materials, Philadelphia, June 1981, pp. 83-85.
[7] Schober, W. and Teindl, H., "Filter Criteria for Geotextiles," *Design Parameters in Geotechnical Engineering,* BGS, London, 1979, Vol. 2, pp. 121-129.
[8] Haliburton, T. A. and Wood, P. D., "Evaluation of the U.S. Army Corps of Engineers Gradient Ratio Test for Geotextile Performance," Second International Conference on Geotextiles, Las Vegas, 1982, Vol. 1, pp. 97-101.
[9] Koerner, R. M. and Ko, F. K., "Long-Term Drainage Capability of Geotextiles," Second International Conference on Geotextiles, Las Vegas, Aug. 1-6, 1982, Industrial Fabrics Association International, St. Paul, MN, 1982, pp. 91-95.
[10] Koerner, R. M., Bove, J. A., and Martin, J. P., "Water and Air Transmissivity of Geotextiles, *International Journal of Geotextiles and Geomembranes,* Vol. 1, No. 1, Elseuire, Essex, England, 1984, pp. 57-74.

L. David Suits[1] and Thomas P. Hoover[2]

Geotextiles and Drainage

REFERENCE: Suits, L. D. and Hoover, T. P., **"Geotextiles and Drainage,"** *Geotextile Testing and the Design Engineer, ASTM STP 952*, J. E. Fluet, Jr., Ed., American Society for Testing and Materials, Philadelphia, 1987, pp. 21-32.

ABSTRACT: The use of geotextiles in geotechnical engineering has brought about many varying theories and approaches. These theories and approaches vary from a broad-based general requirement to a project-by-project design approach.

Provided in this paper are the methods of selection, testing, and the recommended use of geotextiles for drainage applications as defined by the New York State Department of Transportation (NYS DOT) and the California Department of Transportation (Caltrans).

Caltrans approaches the use of geotextiles for drainage in the same manner as they approach aggregate filtration for drainage, with the advantages of a filter with tensile properties being incorporated. The Caltrans Standard Specifications address geotextile use from a range of properties associated with drainage rather than singular absolute parameters.

Caltrans approaches geotextile acceptance on a project-by-project basis. While accepting letters of certification concerning geotextile properties, they also carry on a general control testing program for acceptance.

NYS DOT addresses the use of geotextiles from a preapproved list of materials of five basic categories. This concept provides the basis for NYS DOT use of geotextiles in normal (that is, nonsevere) applications. Therefore, the contractor has a choice of several geotextiles based on availability, economics, and experience in use. The characteristics for the drainage application category are given in the discussion. Also given are the conditions which require specific design considerations in critical applications.

The two points which evolve from the discussion are that both agencies have limited drainage applications to nonwoven materials, and through differing empirical approaches both agencies have developed cost-effective and efficient usage of geotextiles in drainage applications.

KEY WORDS: geotextiles, drainage, permeability, soil retention, woven, nonwoven, apparent opening size, filtration, strength, empirical, Approved List, needle-punched, needle-formed, melt-bonded, permittivity, geotextile testing

In considering geotextiles for usage in drainage applications, the design engineer's primary concern is the removal of free water from the site in question, while retaining the in situ soil in place.

In order to answer this concern, the designer must determine both the permeability and the soil retention characteristics of the geotextile and then compare it to the site conditions when selecting a geotextile.

Since the advent of the use of geotextiles, many different methods of selecting geotextiles have developed. They range from project-to-project design and acceptance to an approved list concept for nonsevere applications requiring a detailed review and design when severe or critical site conditions exist.

The California Department of Transportation (Caltrans) and the New York State Depart-

[1]Soils engineering laboratory supervisor, New York State Department of Transportation, Albany, NY 12232.

[2]Geotechnical engineer, California Department of Transportation, Sacramento, CA 95819.

ment of Transportation (NYS DOT), respectively, represent the ends of the spectrum just described.

The California approach has been to develop a generic specification based on experience and their assumed/developed mechanism of performance. The geotextiles are then accepted on a project-by-project basis.

The New York State approach has been the development of a set of minimum values for permeability, soil retention, and strength characteristics based on early nonsevere experience. If the geotextile in question meets the minimum characteristics, it is placed on an approved list for general use.

The approaches by both agencies are described in detail in following sections, with discussions as to critical points of concern and descriptions of some specific applications.

California Department of Transportation: Geotextile Methods

Generic Specifications for Geotextiles in Drainage

Caltrans uses generic specifications for the geotextiles utilized in all drainage applications. These specifications have been developed and modified throughout our ten years of geotextile use. They are quality control rather than civil engineering specific due to a lack of relevant standard test methods and a lack of understanding of exactly what fabric properties and acceptable limits for those properties are required for effective use of geotextiles. The generic specification system was selected based upon experience and our assumed/developed mechanism of performance.

The assumed performance mechanism, which appears to have been substantiated or confirmed, is "that the geotextile could function as a nondissociative, fine-grained graded aggregate layer, as though it were one layer in a multilayer aggregate filter." Based upon this assumption, Caltrans elected to use geotextiles in a fashion similar to the previously-depended-upon filter aggregate known as "Class 2 Permeable Aggregate." The Class 2 Permeable Aggregate was developed through applied research to provide a suitable filter element for most soils while maintaining high permeability, without requiring site-specific soils information or design. The geotextile performance assumption and our use of Class 2 Permeable Aggregate led to the selection of nonwoven geotextiles for drainage applications. The nonwoven geotextiles, either needle-formed or melt-bonded, were initially picked through a quasitheoretical analysis. That is, geotextile performance characteristics were assumed to be similar to existing filter systems.

First, the performance of woven fabrics was depicted as that of screen, only on a very small scale. Opening sizes and spacings could be measured and compared to the mechanical analysis of the soil to be retained. Using this comparison and the accepted theories, the ratios of pipe perforations to aggregate size, one can make logical choices of geotextiles based upon fabric pore or opening sizes and spacings and the grain sizes of the soil to be filtered. This type of application of geotextiles closely parallels the selection process for the first layer in a graded aggregate filter. It requires extensive soil testing if large areas of fabric-soil contact are required, and extensive knowledge of the available geotextiles. While this offered possible advantages in materials availability, cost, and performance, it still used substantial time and resources for proper design. Additionally, it appeared that only woven filament fabrics would be appropriate for drainage applications. The woven tape or slit-film fabrics did not have sufficient open area and the very fine pore sizes were too widely spaced to provide filtration and maintain high flow capabilities.

Thus, in order to provide more flexibility in choosing fabrics, an investigation of nonwoven fabrics became necessary, requiring an attempt to depict the nonwoven fabrics' characteristics. The first attempt at depicting the characteristics of nonwoven fabrics was to consider them as screens much the same as the woven fabrics. However, the pore size is not constant nor is the spacing. In the melt-bonded fabrics, the spacing and opening sizes are controlled by the random overlaps of the web prior to melt bonding. This results in opening size and spacing ranges. Visual examination of these fabrics indicates that many of the spacings tend to be smaller than the openings. Logically this would be advantageous for drainage filtration applications because it is theoretically impossible to block each and every pore. When one pore is totally blocked, the adjoining pores are kept open by the physical interference of the soil particle blocking the first pore. The pore sizes of the available melt-bonded fabrics seem to be the same order of magnitude as the void size of the Class 2 Permeable Aggregate, thus indicating a possibility for using the melt-bonded fabric in the same manner that the Class 2 permeable Aggregate has been utilized. That is, without a site-specific design. This variable screen analogy does not, however, apply to needle-formed geotextiles.

The needle-formed geotextiles are three dimensional filters, that is, water can run in the plane of the fabric. The other three generic types of geotextiles do not permit water flow within the plane of the fabric. The continuum of void through which the water travels has been termed "tortuous path." This tortuous path precludes the screen analogy used for the other geotextile types. In cross section it also closely approximates the void continuum present in a layer of graded aggregate, except that the solid particles are actually long slender crooked rods. The paths of void continuum are very similar, though somewhat smaller. Thus, the drainage characteristics of fine-grained noncohesive soil were used to depict the drainage characteristics of needle-formed geotextiles. This tortuous path appears to be slightly smaller than the void continuum of Class 2 Permeable Aggregate, thus potentially filtering smaller soil particles.

In addition to the filtering potential, an evaluation of the geotextiles' water-carrying capacity was necessary. This was accomplished through the early stage of Federal Highway Administration (FHWA) Project F76TL07 "Nonwoven Geotextile Fabrics: Evaluation and Specification for Subdrainage Filtration," Translab, Sacramento, California, May 1981. Nonwoven geotextiles were determined to have flow capacities of at least hundreds of cubic metres per day per square metre (thousands of cubic feet per day per square foot) of face contact with water. Having drainage characteristics analogies for the types of geotextiles and suitable high-flow capabilities permitted Caltrans to proceed with some experimental installations. These installations provided an empirical evaluation of the concept of using nonwoven geotextiles in the same manner as Class 2 Permeable Material. The evaluations were empirical because no rigid research control of variables was attempted. The geotextiles were used to solve drainage problems without specific research documentation in an effort to rapidly introduce geotextiles into effective drainage applications. Success or failure was determined by adequate drainage promoting stability or distress elimination. The initial projects were in those situations where a small-scale project needed an immediate solution, where time for soils investigation and filter design was not available, and where consequences of project nonperformance were minimal.

The next class of projects to use geotextiles was in those situations where no other readily available technique could be used to economically provide adequate drainage. Geotextiles were used to solve problems which had no standard solutions available. In retrospect the effective use of geotextiles in most of the early installations was to provide solutions to problems heretofore not economically solvable. Nonwoven geotextiles are currently used in all types of drainage applications.

Applications and Advantages of Nonwoven Geotextiles

Caltrans has used nonwoven geotextiles in a myriad of drainage applications, installing nearly 439 000 m^2 (527 000 yd^2) of fabric in 1984 alone. Applications vary from 0.15-m (6-in.)-wide, 1-m (3-ft)-deep interceptor trenches to 1-m (3-ft)-wide, 8-m (25-ft)-deep cutoff trenches. They include blanket drains, stabilization trenches, and similar type installations. In these applications over the last ten years Caltrans has specified or installed approximately 5 800 000 m^2 (7 000 000 yd^2) of nonwoven geotextiles.

In these applications, geotextiles are simpler to use than graded aggregate, provide a tensile member permitting vertical installation, have proven to be at least as effective as graded aggregate, and are usually more economical.

The nondissociative tensile member permits greater freedom of choice for internal aggregate. Caltrans successfully utilizes permeable aggregates ranging in size from pea gravel to cobbles with and without collector conduits. By using coarser aggregate and letting the geotextile provide the required filtration, lesser volumes of aggregate are necessary. This reduces both import and excavation quantities. Additionally, drainage aggregate is selected based upon one or more of the following: (1) availability; (2) cost; (3) stability (crushed or rounded); (4) permeability and/or handling characteristics, and not based upon filtration characteristics. Separation of sizes in graded aggregate during handling becomes irrelevant, and layer contamination cannot occur.

In one of Caltrans' first applications, geotextiles played a critical role in project development and performance. Drainage was required below an existing box culvert railroad tunnel. Train traffic had to be maintained. And the problem had to be designed, installed, and functioning within six weeks of its discovery, prior to the onset of the wet winter season.

Because the only access was through the floor of the tunnel, numerous impediments were encountered. First was the available space for working between the tracks and tunnel wall, approximately 2.4 m (8 ft). Within this 2.4 m (8 ft), the ballast had to be shored to provide adequate support for daily, sometimes hourly, train traffic. The drainage trench was then excavated around the shoring, box culvert stringers, existing drainage, and railroad utilities. Because of the minimal space and the need to respond to train traffic, the trench was excavated by hand. Also, due to the impediments, the drainage materials used within the trench were installed by hand, usually with multiple handling.

Prior to geotextiles, Caltrans Class 2 Permeable Aggregate would have been used. The separation that occurs from multiple handling would have resulted in compromised filtration. Since the actual water quantity available in winter was unknown and the trench size was dictated by existing physical impediments, it was possible that the Class 2 Permeable Aggregate would not have sufficient permeability to drain the groundwater rapidly enough to solve the problem.

Therefore, Caltrans elected to use a needle-formed fabric with pea gravel and a drainage conduit. The needle-formed fabric was utilized for its flexibility. It readily conforms to variable trench sidewall configurations, thus facilitating hand installation around the numerous impediments. Pea gravel was selected for ease of manual labor handling and the relatively high permeability associated with gap-graded materials. And a conduit was provided to minimize the drainage path within the gravel and to provide a high-volume outlet. This project is approximately nine years old and has performed satisfactorily to date with no water entering the ballast; no fines have piped through the system to the outlet.

Numerous similar trenches have been successful using both melt-bonded and needle-formed fabrics. In fact, geotextiles have been so successful that the Caltrans Standard Specifications for 1984 include a new chapter devoted to geotextiles in drainage and pavement overlay fabrics.

Specification Needs for Utilization Improvement

To quantify geotextile applications, additional specifications are needed that address their geotechnical properties. Quality control specifications are not really substantially different from a list of acceptable products. Both depend on the industry continuing to supply existing fabrics. Tests to evaluate the filtering potential and flow capability per unit area need to be standardized. Tension testing, puncture, creep, and other methods need reevaluation. They need to be of such a scale as to reflect the geotechnical applications, not the textiles industry. Additionally, the test values will need to be approached much the same as soil testing.

For example, permeability of soils is usually considered in orders of magnitude: centimetres/second (feet/day) hundreds of centimetres/second (hundreds of feet/day), etc. This allows for variations of hundreds of percents due to the inherent variabilities of soil, yet provides realistically useful information. Geotextiles should be considered in a similar manner. And, geotextiles should be applied consistent with geotechnical approaches and constraints. Consider flow capability; it should be evaluated and used in terms of order of magnitude, not absolute numbers. The geotextile should be an order of magnitude greater in its water-carrying ability than the soil to be drained and the internal drain rock one or more orders of magnitude greater than the geotextile. As we gain a better understanding of how geotextiles work and improve appropriate quantification techniques, generic specifications should permit using the most effective, most economical products. Until quantification improves, Caltrans will continue the empirical, but highly effective approach, not knowing how conservative our applications are, but taking advantage of the substantial benefits of using nonwoven geotextiles in drainage.

New York State Department of Transportation: Geotextile Methods [*1*]

Background

Over the course of the past 10 to 15 years an evolutionary process has taken place in NYS DOT in the usage of geotextiles. Initially geotextiles were used to solve only special problems, and in situations where had they not worked it would not have been a major problem to revert to conventional means.

After several of these initial applications were successful, NYS DOT was convinced that geotextiles were a viable alternative to conventionally designed solutions for geotechnical problems. That is, they proved to be both an economical and practical solution to the problems encountered.

One such early drainage application success involved the repair of a roadway embankment that was falling into an adjacent stream [*2*]. The solution entailed constructing a berm at the toe of the embankment extending over an area of active silt boils caused by an artesian water condition. The solution to stopping the bubbling of three large silt boils at the embankment toe was to sew sections of a nonwoven geotextile together into a large sheet which was placed over the silt boils and sunk by placing stone directly on the geotextile sheet.

The geotextile properties or characteristics of importance in this application were permeability, soil retention, and strength to withstand installation.

Initially, approval for usage of geotextiles on projects such as this was done on a project-by-project basis. Manufacturers were required to demonstrate that the geotextile being proposed for usage had performed satisfactorily under similar site conditions for the area of proposed usage. Having received approval from the NYS DOT Soil Mechanics Bureau (SMB), the fabric would be accepted in the field on the basis of brand name certification.

As the number and usage of geotextiles increased, the NYS DOT approval procedure evolved to the present Approved List concept. Rather than approve on a project-by-project basis, NYS DOT evaluates a geotextile once, based on manufacturer information and an in-house testing program. If the material meets or exceeds the minimum characteristics to be discussed later in this paper, it is placed on the Approved List in one or more of the five usage categories.

The five categories for approval are listed as underdrain, undercut, slope protection, bedding, and silt fence. Of the five categories, two are considered to be drainage applications using the traditional concept of removing free water from a site, while retaining the in situ soil particles. These categories are underdrain and slope protection.

Underdrain

In the underdrain application, the geotextile is used to line a trench adjacent to a highway pavement which collects free water from underground sources, rainfall, spring melt, etc. Free water is allowed to enter the trench, passing through the geotextile which retains the in situ soil particles. The geotextile prevents clogging and or piping of the underdrain system, the occurrence of which could weaken the subgrade, resulting in damage to the pavement. Figure 1 shows typical underdrain sections, one using a geotextile and one not.

In areas where the soil consists of uniform silts and/or fine sands, piping becomes a problem. The use of a geotextile in place of a multilayer filter is both a practical and economical alternative. The geotextile reduces the chance of piping and prevents the underdrain from becoming clogged.

For this application, the geotextile characteristics which are critical to performance are soil retention and flow rate (that is, permeability). Sufficient strength of the geotextile is only necessary to withstand installation.

In 1975 seven experimental underdrain sections were installed on a project near Ithaca, New York. The purpose of the seven sections was to study combinations of pipe, stone, and various geotextiles. Figure 2 is a plan view of the underdrain system used on this project. The geotextiles used were all nonwoven and varied greatly in their soil retention and flow characteristics. The soils in the area included mixtures of wet gravels, silts, and clays. To date, ten years after installation, the system is performing satisfactorily.

Based on this work, where the trenches were lined with the geotextiles, and the work of others [*3,4*] where the pipe was wrapped with the geotextiles, it is recommended that the procedure of lining the trench with the fabric be used. The basis for this recommendation was finding the filter stone surrounding the wrapped pipe contaminated with soil fines which had migrated in. For long-term installations this could be detrimental to the performance of the system.

Based on the observations of systems using nonwoven and woven geotextiles, only nonwoven geotextiles are approved for usage in underdrains in New York. The observations of nonwoven systems showed little or no silt deposition within the system, whereas those systems using wovens did show deposition.

Slope Protection

In slope protection the geotextile is used as a separator and filter under stone slope protection on highway cut sections. The purpose for the geotextile is to allow the free drainage of groundwater, while retaining the in situ fine soil in place. A stable base on which the stone slope protection can be placed is maintained.

FIG. 1—*Typical sections—underdrain system.*

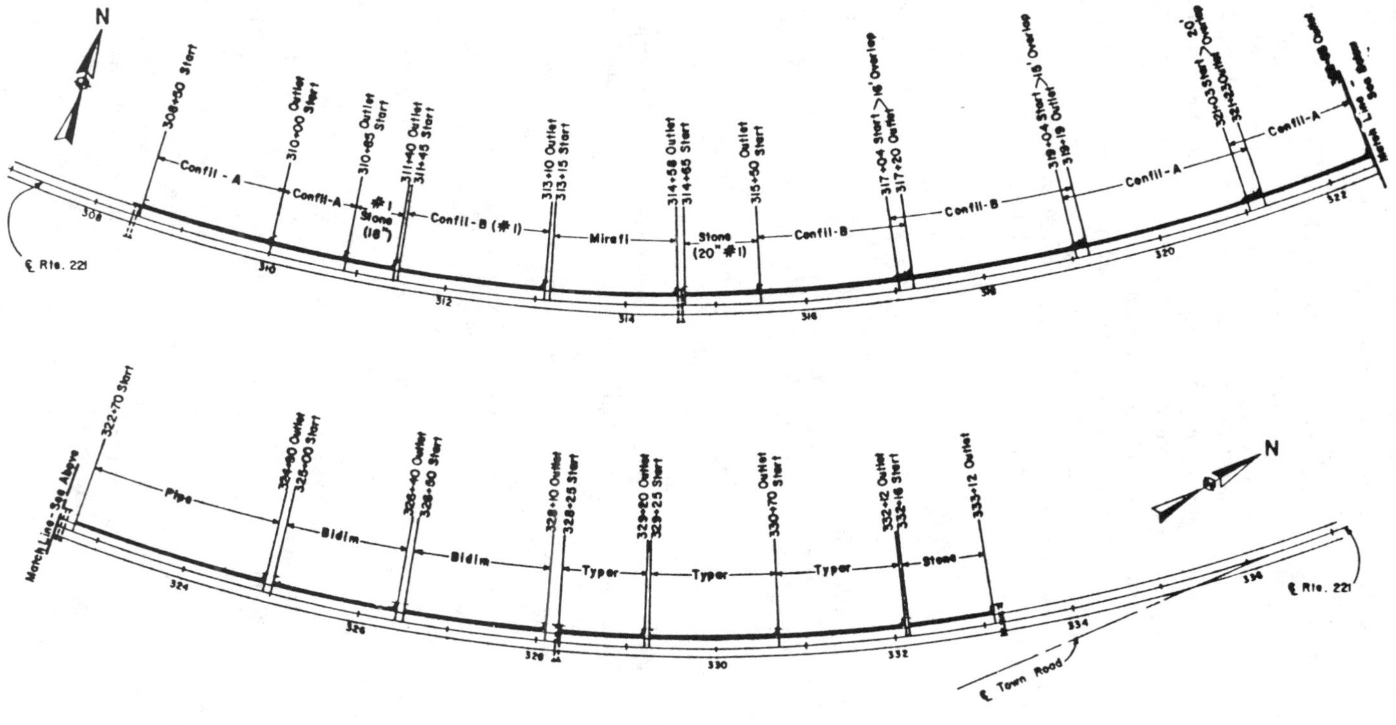

FIG. 2—*Plan scale is 1 in. = 100 feet.*

Surface sloughing of a cut slope where the native soil is either wet silts or fine sands is common. Water emerging from the exposed face of a slope runs down the slope carrying soil with it. The purpose of the geotextile is to allow free drainage while retaining the soil in place, thus preventing sloughing from occurring.

Based on an initial negative experience in this category, NYS DOT now allows only needle-punched, nonwoven geotextiles to be used for slope protection. On one particular project the geotextile was placed on a high cut slope, using a small bulldozer to place a 0.6-m (2-ft)-thick stone blanket over it. Before the installation could be completed the entire mass, fabric and stone, failed by sliding into the ditch. The conclusion reached was that the lack of frictional resistance between the geotextile and the native soil caused the failure. It had been thought that the difference in fabric smoothness would be insignificant.

Based on this experience a third characteristic to be concerned with was added to satisfactory filtration and permeability characteristics for the successful performance of a geotextile in the slope protection category. This third characteristic was satisfactory frictional resistance. This will be discussed in the *Testing and Acceptance* section which follows.

Testing and Acceptance

As discussed earlier, the acceptance procedure for NYS DOT has evolved from a project-by-project basis to an Approved List concept over the past several years.

In order for a geotextile to be placed on the NYS DOT Approved List, the manufacturer/supplier must submit 13 to 17 m^2 (15 to 20 yd^2) of material to the NYS DOT SMB along with a description of the material and, on request, proof of successful usage in the categories of usage being applied for.

The submitted fabric undergoes an in-depth testing program to evaluate the characteristics considered to be critical for satisfactory performance. The categories of testing fall into strength, permeability, and soil retention.

As previously indicated, the strength concern as far as drainage applications is that of surviving installation. If the fabric is torn or punctured during installation, the purpose of allowing water to pass while retaining in situ will be defeated.

With this in mind, a series of four types of strength tests are performed in the approval program. These tests include wide width tension, grab tension, trapezoidal tear, and puncture resistance. The grab tension, trapezoidal tear, and puncture tests have existing ASTM standards, while the wide width tension is a new standard [Test Method for Tensile Properties of Geotextiles by the Wide Width Strip Method (D 4595-86)]. While there are procedural question marks in the existing ASTM standards as applied to geotextiles, they do provide a means of comparing one material to another from the approval procedure.

In the wide width tension test the procedural point of the method of gripping the specimen is presently under study: however, it is felt that, once adopted by ASTM, this test will provide more realistic design values for geotextile strength.

Table 1 lists the minimum values of grab and wide width strength considered acceptable by NYS DOT SMB for the underdrain and slope protection categories.

The permeability characteristic of geotextiles is evaluated in terms of the permittivity of the material. Permittivity is defined as the volumetric flow rate of water per unit area, per unit of head, in the normal direction, through a geotextile, under laminar flow conditions.

The permittivity concept has been adopted because of the difficulty in determining the thickness of a geotextile under test conditions, along with the fact that, due to the varying thickness of geotextiles, Darcy's Coefficient of Permeability can be misleading when comparing one geotextile to another.

The NYS DOT SMB test method, which corresponds to the proposed ASTM method, consists of using a constant head of 50 mm of deaired water to determine the permittivity. The

TABLE 1—*Minimum requirements for geotextiles acceptance.*

Application	Permittivity, s^{-1}	Soil Retention	Min Grab Tension, N(lbf)	Min Wide Width Tension, N/m(lbf/in.)
Underdrain	0.51	Max AOS-297 mm (50 U.S. Sieve)		
		Nonwoven	330 (75)	6484 (35)
Slope protection	0.51	Max AOS-297 mm (50 U.S. Sieve)		
		Needle-punched		
		Nonwoven	352 (80)	6484 (35)

NOTE: AOS = apparent opening size.

proposed ASTM method allows for the usage of either a constant head or a falling head test based on operator experience and preference. Based on round-robin testing done recently and several years ago, it has been determined that when the procedures, as shown in the proposed standard, are followed the constant and falling head methods yield comparable results.

Figure 3 is a picture of the device used by NYS DOT, which meets the requirements established in the proposed ASTM standard.

Table 1 lists the minimum acceptable values of permittivity for the underdrain and slope protection categories.

Darcy's Coefficient of Permeability, as related to permittivity, will be discussed under *Critical Designs*.

FIG. 3—*Permeameter device for permittivity.*

The soil retention characteristics are evaluated through the determination of the apparent opening size or apparent maximum pore diameter within a geotextile.

The apparent opening size of a geotextile is defined as the average diameter, in millimetres (equivalent to a U.S. Standard Mesh Sieve), of that size glass bead for which 5% or more of the beads pass through the interstices of a geotextile.

The method of determination, as proposed by ASTM, consists of sieving various size glass beads, starting with the larger size and progressing to the smaller size, through a single pre-soaked specimen in order to determine the apparent opening size.

In general it is felt that the test is appropriate for woven and thin sheet nonwoven geotextiles. In the thicker nonwovens the glass beads may become trapped within the structure of the geotextile rather than passing through, therefore yielding a questionable value for the test.

Specific values for apparent opening size are shown in Table 1, along with the types of fabric which it is felt demonstrate the best soil retention characteristics.

An important point to emphasize in evaluating the permeability characteristics of a geotextile is the need to collectively examine permittivity and apparent opening size. An impermeable membrane with a single hole punched in it can have the same flow capacity characteristics as a geotextile. However, the soil retention capabilities of the two are quite different. Also, the lower the permittivity, the higher the soil retention capability, and vice versa. Thus, in order to properly select a geotextile, a maximum opening size along with flow rate (permittivity) must be specified. It is up to the designer to determine which characteristic is most important in the specific application and to make any necessary tradeoffs.

In the slope protection discussion, it was pointed out that friction between the geotextile and the native soil was of upmost importance for a successful installation. Presently NYS DOT SMB does not perform any test to determine the frictional characteristics, but has qualitatively established that the needle-punched nonwovens provide the friction behavior necessary.

Table 1 shows the minimum acceptable values for nonsevere or general underdrain and slope protection applications. Critical conditions requiring specific design considerations are discussed under *Critical Designs*.

Once the geotextiles have been tested and shown to meet or exceed the minimum requirements shown in Table 1, the materials are placed on the NYS DOT Approved List. It then becomes the responsibility of the contractor to provide the appropriate material based on availability and economics, with availability being the most critical, so as not to hinder the construction process. The material is then accepted on the job based on brand name certification from manufacturer/supplier that the material is indeed that requested by the contractor.

It is the intention of NYS DOT SMB that once the appropriate ASTM standard test methods are accepted not to have to test geotextiles for general acceptance, but to rely on manufacturer/supplier certification as to meeting the minimum characteristics shown in Table 1. Testing would only take place in critical situations requiring specific design considerations.

Critical Designs

As previously discussed it is necessary to evaluate and determine which characteristic is most critical to the intended application. Based on experience to date NYS DOT feels that in the underdrain application the order of importance is soil retention, permittivity, and strength, with strength being of moderate concern as discussed. In the slope protection category, friction is the most critical, followed by soil retention, permittivity, and strength, with strength again of moderate concern.

It is these characteristics which must be evaluated more thoroughly in the critical application areas as outlined in paragraphs that follow.

With regard to permittivity, it will be necessary to determine a nominal Darcy's Coefficient of Permeability for a geotextile in a critical application in order to determine the geotextile performance in conjunction with the soil in question. This nominal Coefficient can be determined from the permittivity test, using the nominal thickness of the geotextile. The Coefficient of the geotextile must be many times that of the soil in order to insure the ability of the fabric to handle the water flow from the soil.

In the drainage categories, severe conditions are considered to occur when the in situ soil is a fine-to-coarse sand 0.149 to 2.000 mm (No. 100 to No. 10 sieve size), specifically a coefficient of uniformity approaching 1, with a constant source of water and potential high flow. The potential for piping to occur, causing a failure of the system, is high, thus requiring a more in-depth and detailed design process to take place.

In designing under these conditions, it is necessary to consider construction practices along with the specific site conditions. That is, is it possible to install the fabric under the site conditions.

Under these conditions the critical point is to remove the water while retaining the sands. In this light it may be necessary or even unavoidable to allow the passage of silts through the fabric. Allowing buildup of silt deposits in the systems will lead to plugging of the systems and eventual failure.

Without specific site conditions it is difficult to assign ranges of values for permeability and apparent opening size necessary for satisfactory performance. For example, under certain conditions it may be necessary to specify a geotextile having a larger apparent opening size than shown in Table 1. In order to do so a woven material may be necessary.

The important point to remember is that severe conditions as just outlined require more than a cursory review by the design engineer. Specific attention must be paid to both site conditions and construction practices.

General

The NYS DOT SMB approaches the use of geotextiles from an Approved List concept based on manufacturer/supplier certification and an in-depth, in-house testing program. Once it is shown that satisfactory performance has occurred and the values shown in Table 1 have been met, the materials are placed in one or more of five usage categories.

For site conditions considered to be severe, detailed attention must be given to specific site conditions and construction practices or capabilities.

Summary

Both Caltrans and NYS DOT are utilizing geotextiles in drainage applications without benefit of nationally recognized standards for evaluation of the application-relevant fabric characteristics. Through substantially different approaches, both have opted for nonwoven geotextiles in most drainage applications. The Caltrans generic specification is not substantially different than the NYS DOT Approved List approach, since both depend upon the manufacturer continuing to supply the same products as have been previously evaluated. Neither requires a shipment-specific evaluation of the important drainage characteristics, permittivity, and filtration or plugging potentials. Adoption of the proposed ASTM Test Methods for geotextiles should help remedy this situation.

The lack of national standards for evaluating geotextiles has not precluded their effective use. Both NYS DOT and Caltrans have successfully utilized geotextiles in drainage applications for about ten years. The applications range from relatively minor trenches to large stabilization drainage trenches. The tensile properties associated with geotextiles and their ability to readily conform to irregular soil surfaces, lacking soils in place, results in their being much easier to place than are graded aggregate systems. Additionally, they are usually more economical than graded aggregate systems, while providing comparable filtration and maintaining high water drainage capacities.

Geotextiles are a new civil engineering drainage tool that is here to stay. Their utilization, while not readily quantifiable at this time, is not complex. Nonwoven geotextiles can provide filtration and maintain permeability without site-specific design. Both job-specific specification and acceptance and acceptable products lists are viable approaches to utilizing geotextiles in drainage. Caltrans and NYS DOT have recognized the benefits of using geotextiles and through different approaches have incorporated them into their standard methods of providing effective subsurface drainage.

References

[1] Minnitti, A., Suits, L. D., and Dickson, T. H., *New York State Department of Transportation's Experience and Guidelines for the Use of Geotextiles,* 62nd Annual Meeting, Transportation Research Board, Washington, DC, Jan. 1983.

[2] Carlo, T. A., *"Embankment Stabilization Utilizing a Non-Woven Filter Cloth,"* at Route 55, 55A, Napanoch-Montela, Ulster County, New York, Soil Mechanics Bureau, New York State Department of Transportation, Albany, NY, Dec. 1977.

[3] Forshey, A. D., "Title 76-16: Use of Filter Cloths for Subsurface Drainage Systems for Highways," proposal prepared for the FHWA by Commonwealth of Pennsylvania Department of Transportation, Bureau of Materials Testing and Research, Harrisburg, 18 Aug. 1976.

[4] Forshey, A. D., "Use of Filter Cloths for Subsurface Drainage for Highways," *Highway Focus,* Vol. 9, No. 1, May 1977, pp. 82–87.

Robert M. Koerner[1] *and John A. Bove*[2]

Lateral Drainage Designs Using Geotextiles and Geocomposites

REFERENCE: Koerner, R. M. and Bove, J. A., **"Lateral Drainage Designs Using Geotextiles and Geocomposites,"** *Geotextile Testing and the Design Engineer, ASTM STP 952,* J. E. Fluet, Jr., Ed., American Society for Testing and Materials, Philadelphia, 1987, pp. 33–44.

ABSTRACT: Of the available functions that geosynthetic materials possess, their ability to conduct liquids in the plane of their structure is important in a wide class of drainage applications. The thicker geotextiles (usually needled nonwoven types) and a new generation of high drainage composite materials are most suitable for this purpose.

This design-oriented paper gives the necessary theory and required test data and then proceeds to solve three numeric examples illustrating the procedures. They are slope drainage, consolidating foundation drainage, and drainage behind a retaining wall.

KEY WORDS: geotextiles, geocomposites, geosynthetics, design, drainage, transmissivity, flow rate, planar flow

Over the relatively short history of geotextiles, three distinct design approaches have emerged. They are design by price, design by specification, and design by function [*1*]. Design by price recognizes that fabric properties are related to fabric mass per unit area (weight) and in turn to cost per unit area. Thus, one purchases the "best" fabric for available funds. Of course, this very simplistic method is often dangerous, but when tempered with sufficient experience can serve (and has served) the industry well. Much more widely practiced (particularly by governmental agencies) is design by specification. In this method common geotextile application areas are paired with limiting (for example, minimum average roll values) geotextile properties in the physical, mechanical, hydraulic, and/or endurance areas. If the candidate geotextile has properties meeting, or exceeding, the listed values, it is acceptable. If more than one geotextile is acceptable, the decision is then made on the basis of cost and availability. The third method, design by function, recognizes that a primary function (depending upon the particular application) can generally be identified. This being the case, the relevant fabric properties relating to this primary function are quantitatively determined and then compared to the candidate geotextile's actual properties. The ratio of these values (actual to required) is the factor of safety. This method is essentially a case-by-case design procedure but one which is very important for large and/or critical applications. It is the method that will be utilized in this paper.

When considering "design by function" one must first assess the application in light of the following available functions:

1. Separation.
2. Reinforcement.
3. Filtration.
4. Drainage.
5. Moisture barrier.

[1]Professor of Civil Engineering, Drexel University, Philadelphia, PA 19104.
[2]Project engineer, Soil and Material Engineers, Inc., Fairfield, OH 45014.

This paper focuses entirely on the drainage function. Here liquid (usually water) is to be conveyed in the plane of the fabric structure. Note that this is *not* the cross plane fabric function of filtration which has its own separate (although in many respects similar) design methodology [*2*].

The application areas where drainage within the plane of the fabric structure is important are numerous. Some are listed below:

1. As a chimney drain in an earth dam.
2. As a drainage gallery in an earth dam.
3. As a drainage interceptor for horizontal flow.
4. As a drainage blanket beneath a surcharge fill.
5. As a drain behind a retaining wall.
6. As a drain beneath railroad ballast.
7. As a water drain beneath geomembranes.
8. As a gas drain beneath geomembranes.
9. As a drain beneath sport and athletic fields.
10. As a drain for roof gardens.
11. As a pore water dissipator in earth fills.
12. As a replacement for sand drains.
13. As a capillary break in frost sensitive areas.
14. As a capillary break for salt migration in arid areas.
15. To dissipate seepage water from exposed soil or rock surfaces.

This paper will present the necessary hydraulic considerations needed for drainage design and comment upon the required data base. It will then illustrate the design process for a number of practical applications using both geotextiles and drainage geocomposites.

Hydraulic Considerations

When the flow within a fabric structure is saturated and laminar, Darcy's formula can be used in drainage design to great advantage. Consider the general situation as shown in Fig. 1.

$$q = k_p i A$$

$$= k_p \frac{\Delta h}{L} (W \times t)$$

$$= (k_p t) \frac{\Delta h \times W}{L}$$

$$k_p t \equiv \theta = q \frac{L}{\Delta h \times W}$$

where

q = flow rate,
k_p = planar coefficient of permeability,
Δh = total head causing flow,
W = width,
L = length,
t = thickness,
θ = transmissivity,

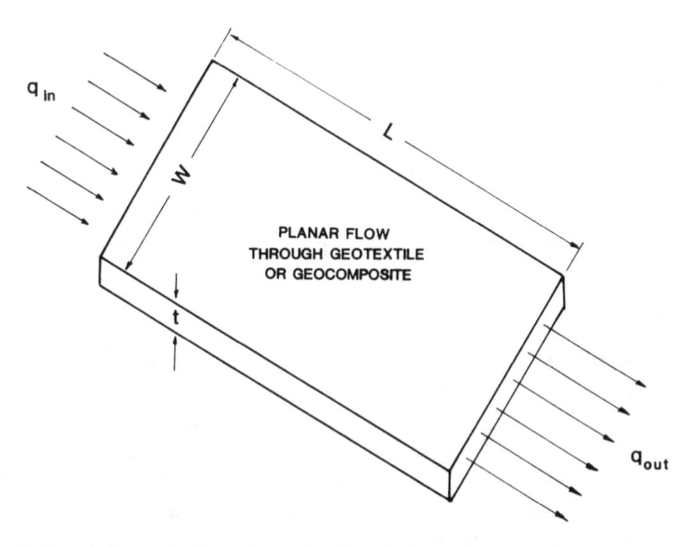

FIG. 1—*General schematic for in-plane drainage in geosynthetic materials.*

i = hydraulic gradient, and
A = cross-sectional area.

The only uniqueness of this development over traditional geotechnical engineering work is the necessity of defining the term called *transmissivity*, "θ." As indicated, it is the permeability coefficient times the thickness. This is convenient due to the compressibility of many fabric systems used in drainage applications at varying normal stresses. The design approach is to determine the required transmissivity (θ_{reqd}) from design charts, field measurements, flow net analysis, etc., and compare it to the laboratory determined value of the candidate geotextile's transmissivity (θ_{act}), thereby forming a factor of safety.

$$FS = \theta_{act}/\theta_{reqd}$$

when

FS = factor of safety,
θ_{act} = actual transmissivity, and
θ_{reqd} = required transmissivity.

Depending upon the criticality of the situation, this factor of safety should range from 2.0 to 5.0.

In addition to conventional geotextiles, there are available a number of composite materials (which will be referred to as drainage geocomposites, or simply geocomposites) which have very high drainage capability. These composite materials often consist of an open core material covered by a geotextile filter. In most of these geocomposites, flow will not be saturated nor laminar. If either situation is the case, Darcy's formula cannot be used. In such cases, the flow rate must be compared directly, that is

$$FS = q_{act}/q_{reqd}$$

when

FS = factor of safety,
q_{act} = actual flow rate, and
q_{reqd} = required flow rate.

It should be recognized that some geocomposite core materials are very compressible. Even hand pressure can measurably deform them. Thus the data base needed to obtain q_{act} in the just-cited equation must reflect the normal pressure to be exerted on the material in the field.

Required Data Base

The required data base for determining the transmissivity (θ_{act}) or flow rate (q_{act}) of geotextiles and geocomposites comes from laboratory measurements on the materials in question. Work of this type has been published for geotextiles [*3–6*] and is further reviewed in this publication by Carroll [*7*]. A standardized test method is currently under review by ASTM Committee D35 on Geotextiles, Geomembranes, and Related Products. Using the data of Gerry and Raymond [*4*], Table 1 illustrates that of the different types of fabrics only the nonwoven-needled geotextiles have appreciable in-plane flow capability. Thus this type of geotextile is preferred for in-plane drainage applications. Koerner and Bove [*5*] have tested a number of commercially available nonwoven-needled fabrics with the results shown in Fig. 2. Here a number of observations can be noted:

1. All fabrics show an exponentially decreasing trend due to initial compression of these lofty fabrics at low stresses.
2. All fabrics show a nearly constant transmissivity value at stresses higher than approximately 19 kPa (400 psf) where the fiber structure is sufficiently dense to support the applied stress.
3. This constant, and residual, value is in the range of 0.40 to 1.4×10^{-6} m^3/s-m (0.003 to 0.010 ft^3/min-ft), which is the value to be generally used in design.
4. There is considerable crossover of trends in the data from the various geotextiles that were tested.
5. There is, however, a general trend that the heavier and/or thicker geotextiles have the highest transmissivity.

The situation with geocomposites is considerably different. A number of these materials, specifically directed at high in-plane drainage, are listed in Table 2. They each have their own

TABLE 1—*Typical values of drainage capability (in-plane flow) of geotextiles,*[a] *data after Gerry and Raymond* [4].

Type of Geotextile	Transmissivity[b]		Permeability Coefficient	
	m^3/s-m	ft^3/min-ft	cm/s	ft/min
nonwoven-heat set	3.0×10^{-9}	2.0×10^{-6}	0.0006	0.0012
woven-slit film	1.2×10^{-8}	8.1×10^{-6}	0.002	0.0039
woven-monofilament	3.0×10^{-8}	2.0×10^{-6}	0.004	0.0079
nonwoven-needled	2.0×10^{-6}	1.3×10^{-3}	0.04	0.079

[a]Values taken at applied normal pressure of 40 kPa (830 psf).
[b]Transmissivity, $\theta = k_p t$, where k_p is the planar permeability coefficient and t is the fabric thickness at the applied pressure.

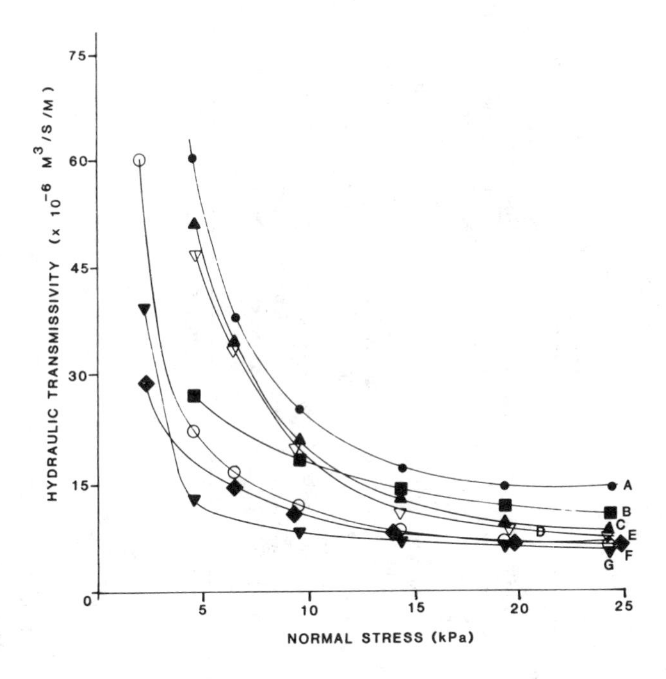

Geotextile	Mass Per Unit Area		Nominal Thickness		Polymer	
	oz/yd²	g/cm²	Mils	mm	Type	Filament
A	16	540	210	5.3	PET	continuous
B	18	600	190	4.7	PET	staple
C	18	600	150	3.8	PP	continuous
D	16	540	160	4.1	PP	continuous
E	12	400	110	2.8	PP	continuous
F	14	470	130	3.3	PP	staple
G	16	540	110	2.8	PET	continuous

FIG. 2—*Transmissivity response versus applied normal stress for various needled nonwoven geotextiles, after Koerner and Bove* [5].

unique features with the core compressibility under load being a major consideration. Compare, for example, the flow rate response from a stiff core material like Cordrain, shown in Fig. 3, with that of a compressible core material like Enkadrain, shown in Fig. 4. These types of response curves are necessary for the designs to follow. Their testing methods and behavior are described by Carroll [7].

Design Using Bulky Geotextiles

In this section, two separate designs will be presented: one is gravity drainage using a geotextile against a seeping soil slope; the other is pressure drainage for use of a geotextile under a soil surcharge fill.

TABLE 2—*Summary of geocomposite drainage systems.*

Trade Name	Manufacturer	Core Characteristics				Geotextile Filter
		Thickness		Shape	Material	
		mm	Inches			
Enkadrain	American Enka Corp.	10 and 20	0.4 and 0.8	Wire web	Nylon	Nonwoven-needled
Miradrain	Mirafi Corp.	25	1.0	Waffle or columns	Polystyrene	Nonwoven-heat set or needled
Cordrain	Burcan Manufacturing Co.	25	1.0	Waffle	Polystyrene	Nonwoven-heat set
Geotech	Geotech Systems Corp.	51	2.0	Beads with bitumen binder	10 mm (3/8 in.) expanded polystyrene	Nonwoven-needled
Filtram	ICI Fibers	5	0.2	8 mm (0.3 in.) square grid	Polypropylene	Nonwoven-heat set
Amerdrain	International Construction Equipment	8	0.32	6 mm (0.25 in.) corrugated ribs	Polypropylene/ polyethylene	Nonwoven-heat set
Hydraway	Monsanto	25	1.00	6 mm (0.25 in.) cylinders	Polyethylene	Nonwoven-needled

NOTE: Addresses and phone numbers of companies:

American Enka Corp., Enka, NC 28728; (704) 667-7713.
Mirafi, Inc., P.O. Box 240967, Charlotte, NC 28224; (800) 438-1855.
Burcan Mfgr. Inc., 111 Industrial Drive, Whitby, Ontario, Canada L1N5Z9; (416) 668-3131.
Geotech Systems Corp., 1516 Spring Hill Rd., McLean, VA 22101; (703) 893-1310.
ICI Fibres, "Terram," Ponty Pool, Gwent, NP48Yd Great Britain; (04955) 58150.
International Construction Equipment, Inc. (American Wick Drain Co.), 301 Warehouse Drive, Matthews NC, 28105; (800) 438-9281.
Monsanto Engineered Products Division, 800 N. Linbergh Blvd., St. Louis, MO 63166; (314) 694-1000.

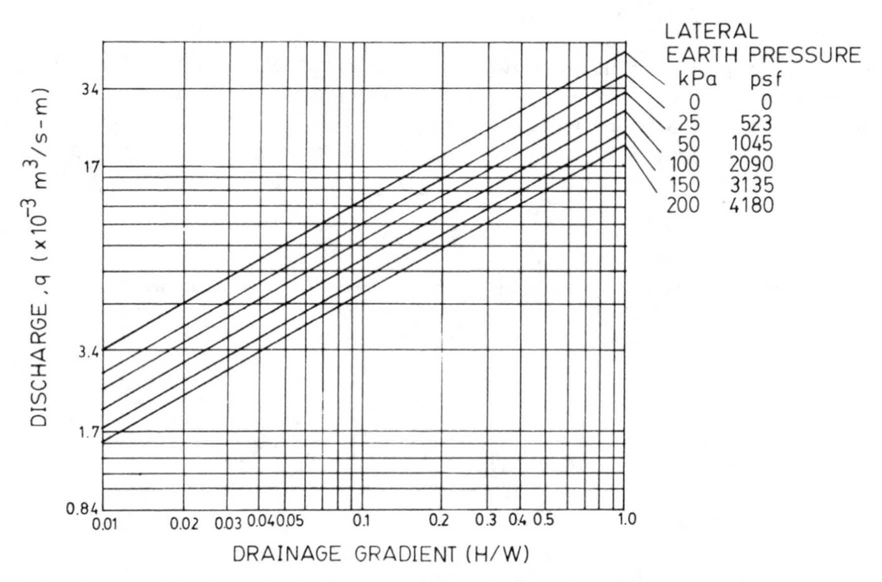

FIG. 3—*Flow rate behavior of Cordrain at hydraulic gradients of 0.01 to 1.0, after Burcan Manufacturing Co.*

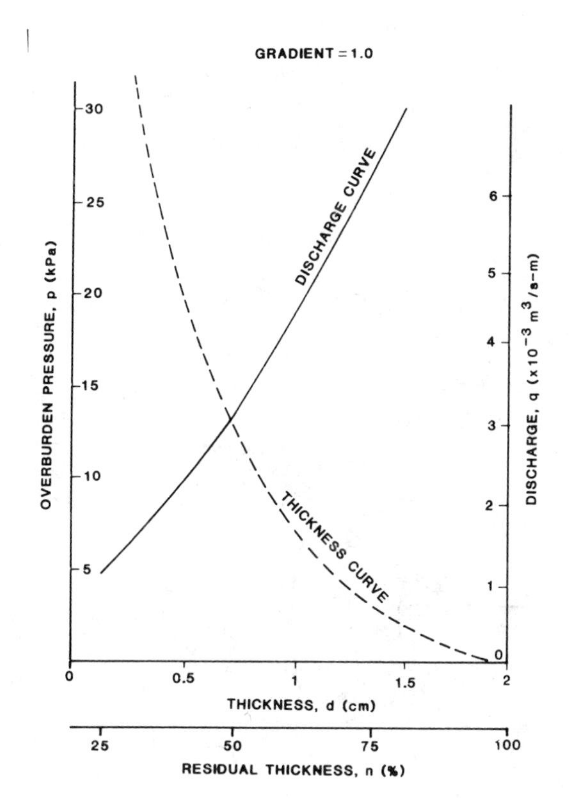

FIG. 4—*Flow rate behavior of Enkadrain at hydraulic gradient of 1.0, after Enka Industrial Systems.*

Example No. 1

Given a 4-m-high earth embankment whose seepage rate can be obtained by the flow net shown in Fig. 5. Since the slope will eventually be covered by a liner (maximum stress of 20 kPa), it is desired to drain it with a nonwoven needle-punched geotextile whose transmissivity response is shown in Fig. 2 as Curve "C." What is the Factor of Safety of this particular geotextile against the anticipated flow?

Solution

(a) Determine the flow rate coming to the geotextile from flow net theory.

$$q = khF/N$$

$$= (1.2 \times 10^{-6})(3)(4/6)$$

$$= 2.4 \times 10^{-6} \text{ m}^3/\text{s-m}$$

(b) Calculate the hydraulic gradient.

$$\sin \beta = \sin 35° = 0.57$$

(c) Calculate the required transmissivity.

$$q = k_p i A$$

$$= k_p i (t \times W)$$

$$\theta_{\text{reqd}} = k_p t = q/(i \times W)$$

$$= \frac{2.4 \times 10^{-6}}{(0.57)(1.0)}$$

$$= 4.2 \times 10^{-6} \text{ m}^3/\text{s-m}$$

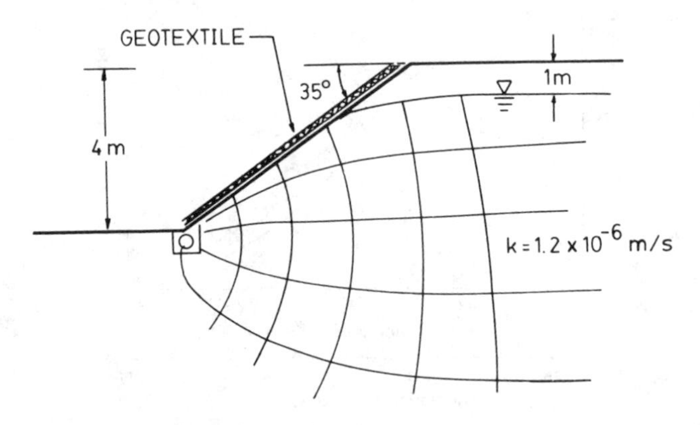

FIG. 5—*Example problem sketch for gravity flow into a geotextile drain on a sloped embankment.*

(d) Obtain the actual transmissivity of the candidate geotextile at the stress anticipated—from Fig. 2, Curve "C."

$$\theta_{act} = 10 \times 10^{-6} \text{ m}^3/\text{s-m}$$

(e) Calculate and assess the resulting factor of safety.

$$\text{FS} = \theta_{act}/\theta_{reqd}$$

$$= 10 \times 10^{-6}/4.2 \times 10^{-6}$$

$$\text{FS} = 2.4$$

Example No. 2

Given a 15-m-wide surcharge fill, as shown in Fig. 6, which is placed in twelve days on a silty clay foundation soil of permeability, $k_{soil} = 2.1 \times 10^{-8}$ m/s and coefficient of consolidation, $c_v = 8.8$ m^2/year. Will the geotextile labeled "A" in Fig. 2 be adequate to drain the expelled foundation water without it entering into the surcharge fill?

Solution

(a) Determine the required transmissivity using the theory developed by Giroud [8].

$$\theta_{reqd} = \frac{B^2 k_{soil}}{(c_v t)^{1/2}}$$

where

B = width of surcharge fill,
k_{soil} = soil permeability,
c_v = coefficient of consolidation of soil, and
t = time for surcharge fill to be placed.

$$\theta_{reqd} = \frac{(15)^2\,(2.1 \times 10^{-8})}{[(8.8)(12/365)]^{1/2}} = 8.8 \times 10^{-6} \text{ m}^3/\text{s}$$

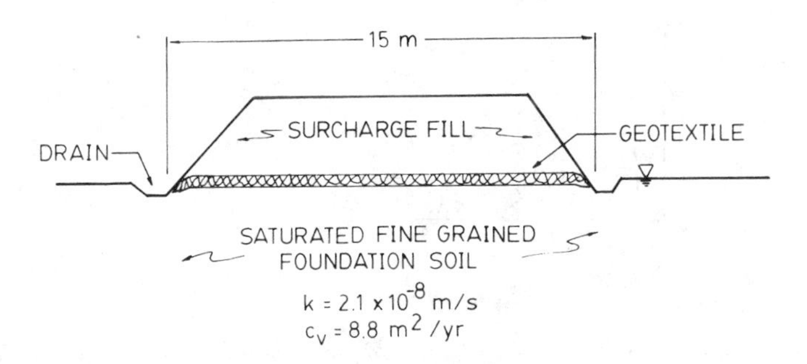

FIG. 6—*Example problem sketch for pressure flow into a geotextile drain under a surcharge fill.*

(b) Compare this to the candidate geotextile's transmissivity from Fig. 2 to calculate the actual factor of safety.

$$\text{FS} = \theta_{act}/\theta_{reqd}$$

$$= (15 \times 10^{-6})/(8.8 \times 10^{-6})$$

$$= 1.7$$ too low, either use a thicker geotextile, multiple layers, or a geocomposite.

Design Using Geocomposites

With geocomposites the flow regime is quite complex. Most of those listed in Table 2 can handle 0.023 m^3/s-m (30 L/min-m or 2.0 gal/min-ft), which strongly suggests turbulent flow. Hence transmissivity values originated from Darcy's formula cannot be used and flow rates themselves must be compared. The following example illustrates the technique for both "stiff" and "flexible" geocomposites.

Example No. 3

For the geocomposite drain placed behind the cantilever retaining wall shown in Fig. 7, calculate the flow factor of safety based on Cordrain (see Fig. 3) and Enkadrain (see Fig. 4). (Note that a similar problem is examined in Ref *2* where 13 layers of geotextiles would have been required.) The related soil properties are:

$$k_{soil} = 5 \times 10^{-5} \text{ m/s}$$

$$\text{Unit weight, } \gamma = 16.5 \text{ kN/m}^3$$

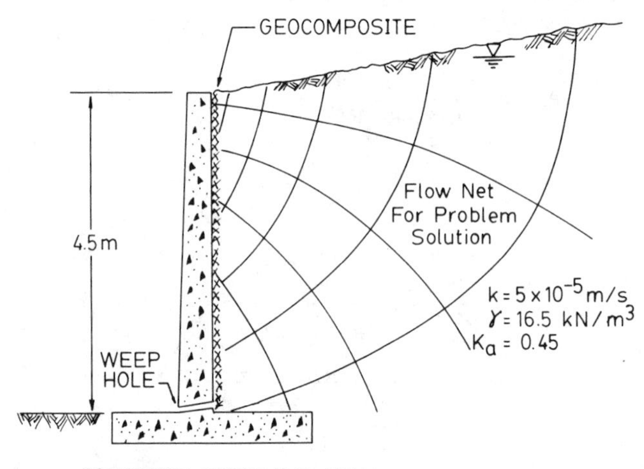

FIG. 7—*Example problem sketch for drainage into geocomposites used as a vertical drain behind a retaining wall.*

A full head should be assumed. The ratio of horizontal to vertical soil pressure (K_a) at the base of the wall is 0.45.

Solution

(a) Draw the flow net (as shown in Fig. 7) from which the flow rate is calculated.

$$q = khF/N$$

$$= (5 \times 10^{-5})(4.5)(5/5)$$

$$= 2.25 \times 10^{-4} \text{ m}^3/\text{s-m}$$

(b) Calculate the hydraulic gradient which, for a vertical application, is sin 90° = 1.0.

(c) Calculate the maximum pressure exerted by the soil on the geocomposite. In this case, it will be at the bottom of the wall stem section and is based on soil pressure alone. If an unbalanced hydrostatic pressure also exists, it must also be included. This would be a site specific situation.

$$\sigma_n = K_a \gamma z$$

$$= 0.45\ (16.5)\ 4.5$$

$$= 33.4 \text{ kN/m}^2$$

$$\cong 33 \text{ kPa}$$

(d) Obtain the actual flow rate of the candidate geocomposites at the gradient and normal stress as calculated and determine the factor of safety.

For Cordrain (from Fig. 3)

$$q_{\text{act}} = 3.5 \times 10^{-3} \text{ m}^3/\text{s-m}$$

$$\text{FS} = \frac{q_{\text{act}}}{q_{\text{reqd}}}$$

$$= \frac{3.5 \times 10^{-3}}{2.25 \times 10^{-4}}$$

$$= 15.6 \text{ OK}$$

For Enkadrain (from Fig. 4)

$$q_{\text{act}} = 1.5 \times 10^{-3} \text{ m}^3/\text{s-m}$$

$$\text{FS} = \frac{q_{\text{act}}}{q_{\text{reqd}}}$$

$$= \frac{1.5 \times 10^{-3}}{2.25 \times 10^{-4}}$$

$$= 6.7 \text{ OK}$$

Conclusions

Presented in this paper were designs involving geotextiles and geocomposites used for in-plane flow applications. They illustrated that the drainage function is well suited for bulky geotextiles and the more recently introduced drainage geocomposites. Furthermore, the problems illustrated that this latter group of materials (reviewed in Table 2) is capable of handling very large flow volumes within their respective core structures. All of the problems presented, however, could not have been quantitatively approached without the necessary data base. This data base must be laboratory generated, a task to which ASTM Committee D35 (and other standards groups around the world) is currently undertaking. Within this data base is the necessary flow rate capability of the drainage materials involved versus hydraulic gradient at the applied normal stress it will be functioning at. Beyond this the calculations proceed as with standard geotechnical engineering drainage design problems. No more, nor no less, information is needed than with design without geotextiles. Proceeding in this way, rational designs to a wide range of drainage problems can be solved. At this point "design with confidence" is within our grasp.

References

[1] Koerner, R. M., "A Note on Geotextile Design Methods," Geotechnical Fabrics Report, International Fabrics Association International, St. Paul, MN, Vol. 2, No. 2, Fall 1984, pp. 28-29.

[2] Koerner, R. M., "Geosynthetics and Their Use in Filtration and Drainage Applications," in *Proceedings,* Use of Geotextiles, Geogrids and Geomembranes in Engineering Practice, 28 Nov. 1984, Toronto, Canadian Geotechnical Society, Southern Ontario Section, Ottawa, Canada, pp. 1-16.

[3] Koerner, R. M. and Sankey, J. E., "Transmissivity of Geotextiles and Geotextile/Soil Systems," *Proceedings,* Second International Conference on Geotextiles, 1-6 Aug. 1982, Las Vegas, NV, pp. 173-176.

[4] Gerry, B. S. and Raymond, G. P., "The In-Plane Permeability of Geotextiles," *Geotechnical Testing Journal,* Vol. 6, No. 4, Dec. 1983, ASTM, Philadelphia, pp. 181-189.

[5] Koerner, R. M. and Bove, J. A., "In Plane Hydraulic Properties of Geotextiles," *Geotechnical Testing Journal,* Vol. 6, No. 4, Dec. 1983, ASTM, Philadelphia, pp. 190-195.

[6] Koerner, R. M., Bove, J. A., and Martin, J. P., "Water and Air Transmissivity of Geotextiles," *Journal of Geotextiles and Geomembranes,* Vol. 1, 1984, pp. 57-73.

[7] Carroll, R. B., "Hydraulic Properties of Geotextiles," this publication.

[8] Giroud, J.-P., "Designing with Geotextiles," *Materials and Structures,* RILEM, Bordas-Dunod Publications, Paris, 1981, pp. 257-272.

Ronald K. Frobel,[1] *Gerhard Werner*,[2] *and Manfred Wewerka*[2]

Geotextiles as Filters in Erosion Control

REFERENCE: Frobel, R. K., Werner, G., and Wewerka, M., **"Geotextiles as Filters in Erosion Control,"** *Geotextile Testing and the Design Engineer, ASTM STP 952,* J. E. Fluet, Jr., Ed., American Society for Testing and Materials, Philadelphia, 1987, pp. 45-54.

ABSTRACT: Geotextiles are often used in place of conventional mineral filters in erosion control applications along lake or ocean shorelines, canals, stream channels, and other hydraulic structures. The basic concepts of a geotextile filter system are explained, and basic applications are described. The functional design considerations for mechanically needle-punched nonwoven geotextiles are discussed with reference to current filter criteria, and filter design based on known values of soil particle size and uniformity coefficients is presented in relatively simple graphical form. In particular, filter design criteria for laminar monodirectional flow and granular soil and turbulent, alternating water flow and granular soil are presented.

KEY WORDS: geotextiles, fabrics, erosion control, filters, granular soil, filter criteria

Five basic end-use functions are performed by geotextiles when used in the design and construction of earth structures, pavement systems, and other manmade structures supported on or covered by earth/geotextile systems. The functions include the following:

1. *Separation*—Wherein the geotextile forms a boundary between different soil or rock materials, thus segregating two or more particle sizes.
2. *Reinforcement*—The geotextile imparts tensile strength to an earth/geotextile system, thereby increasing structural stability.
3. *Filtration*—The geotextile effectively retains particles while allowing water to flow through with little or no increase in pore pressure to the surrounding soil.
4. *Drainage*—The geotextile effectively allows water to flow in the plane of the fabric, thus allowing drainage of water away from a structure or system.
5. *Moisture Barrier*—The geotextile when coated or impregnated with a relatively impermeable material such as asphalt effectively forms a barrier which impedes the flow of moisture through a system.

All those basic or primary functions just stated also serve as secondary functions to each other in given applications. In erosion control and in particular erosion control for coastal and riverbank protection, the primary function of a geotextile is filtration with the separation function playing a secondary role. A tertiary function in some erosion control applications would be reinforcement.

Concept

Embankments along lake or ocean shorelines, inland waterways, canals, artificial stream channels, and other hydraulic structures are continually subjected to wave and current action as

[1]Technical manager, Chemie Linz U.S., Inc., Golden, CO 80401.
[2]Technical manager—Geotextiles and technical support engineer, respectively, Chemie Linz AG, Linz, Austria.

well as fluctuating water levels. These conditions can cause severe slope erosion and instability if the slope is not protected. Conventional armor protection such as concrete slabs and blocks and riprap can effectively protect the slope. However, water can still cause severe erosion if it is allowed to contact the underlying supporting soil, resulting in the undermining of the armor's foundation. Traditional bedding or filter materials placed between the armor and the underlying soil have consisted of granular filters, specially graded sand, gravel, or stone. However, these materials can also be subject to erosive forces that will eventually wash out the bedding material and eventually cause exposure of the embankment soils and potential destruction of the armor protection.

The original use of geotextiles in the United States in the early 1960s was as a replacement for the conventional aggregate filter used under armor-type shore protection Figure 1 shows the basic cross section of both a conventional aggregate filter and the replacement geotextile filter system.

The geotextile can effectively replace the aggregate filter system or replace one of the layers in a multilayer filter system. A geotextile in combination with aggregate riprap or gravel can also be used to reduce surface erosion due to internal embankment piping or surface runoff.

Both woven and nonwoven geotextiles have been used in erosion control applications. Regardless of the type or structure of the fabric, the geotextile must withstand installation stresses and thus possess high strength survivability characteristics such as relatively high tear and puncture resistance as well as in-service tensile stresses and filtration and environmental durability criteria.

Basic Applications

Erosion protection applications can be used in the construction of the following:

1. Revetments or seawalls.
2. Cut and fill slopes.
3. Ditch, channel, and canal slope protection.
4. Stream bank protection.
5. Scour protection around bridge piers or other structures located in a hydraulic channel or waterway.

Revetments or seawalls are constructed with stone or concrete armoring with the primary purpose of protecting the shore of a large lake or an ocean beachfront from wave action and subsequent washing out and loss of material. With respect to maritime installations, thought must be given to rapid tidal changes as well, which could cause rapid drawdown and erosion of the fine material under the protecting armor. Figure 2 illustrates a typical armored revetment. When using large stones or precast concrete segments, consideration must also be given to the strength of the geotextile during installation, especially the puncture resistance, tensile strength, and elongation (ability to conform to changing subsoil under load within stress limitations).

Cut and fill slopes in highway, building, and canal construction are usually designed on the basis of stability of slope material and economics of cut/fill quantities. Consideration must also be given to the erodibility of the soil due to surface runoff (from rain) or groundwater seepage. If the surface soil is erodible, then surface protection must be utilized. Surface protection can be vegetation growth, benching or heavier gravel, or aggregate cover. Ditches and channels must also be protected from surface runoff erosion, but more importantly, they must be protected from the flowing water within the hydraulic conveyance structure. In this case, maximum flow rates and velocities must be known so as to design the protective select gravel or aggregate erosion protection. This is also true for natural stream bank protection, especially at abrupt

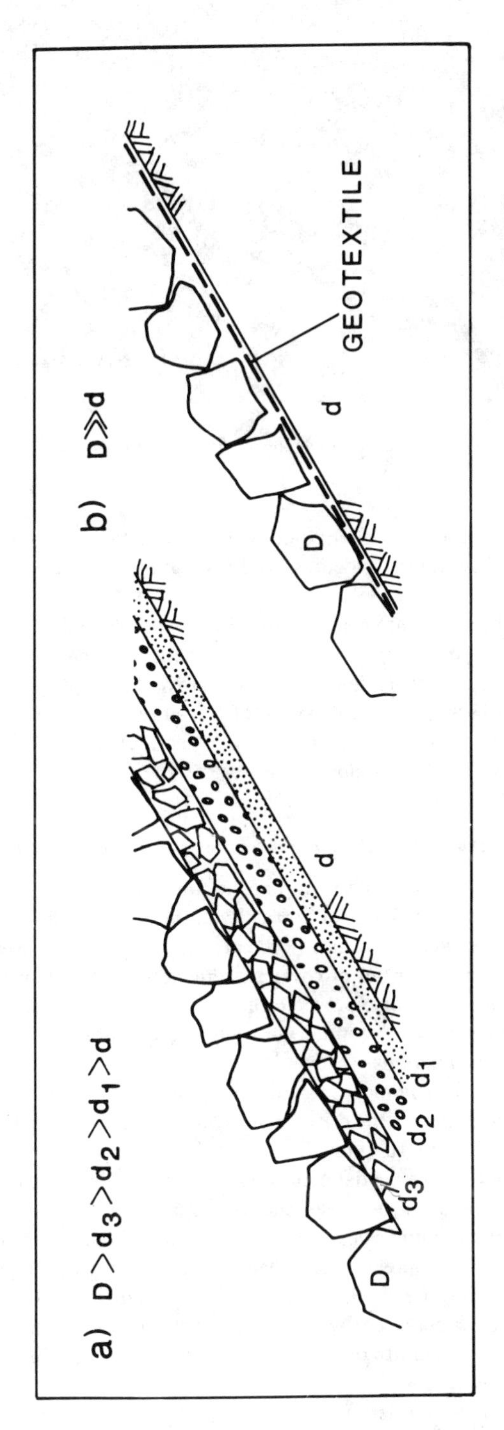

FIG. 1—*Filter systems under riprap slope protection: (a) graded aggregate filter system; (b) geotextile filter system. Where* D = *diameter of armor and* d = *diameter of mineral filter or subsoil.*

FIG. 2—*Armored revetment with geotextile filter.*

changes in stream direction and subsequent bank scour. Stream bank erosion protection can be more critical in that the flow rates, velocities, and turbulence can be extreme in high runoff areas.

Within hydraulic conveyance channels, streams, and rivers, as well as the sea, can be found numerous types of concrete, stone, wood, or steel projections forming piers or supporting abutments for bridge crossings and other manmade structures. These projections from the river, channel, or sea floor can be subjected to erosion of the supporting or surrounding soil if not protected from scour. Scour protection can be a very critical design consideration for any water crossing that demands support from below the water surface.

Functional Design Considerations in Utilizing a Geotextile as a Filter in Erosion Control

In considering the use of a geotextile to replace a conventional granular filter, the design engineer must evaluate the various kinds of fabrics available and the filtration behavior associated with the type of fabric structure. In addition, care must be taken to select a fabric that will not puncture under conventional armor or cover placement as well as conform to the changing subsoil conditions under load and also preserve the filtering efficiency under loading. Experience has shown that the mechanically bonded continuous filament spunbonded nonwovens can perform all of the just-cited functions.

Mechanically needle-punched nonwoven geotextiles form a three-dimensional structure of fibers. It is this feature which makes the difference in filtration performance compared with the woven or gridlike planar fabric structures. Theoretically, it is the fiber distance distribution along the cross section that allows partial penetration of the soil into the nonwoven geotextile. With increasing penetration, the structure of the soil apparently becomes loftier, the particles together with the fibers thus forming a coarse filter at the points of sedimentation. This penetration property or bridging effect increases steadily outward, starting at the geotextile/soil interface, building a filter system quite similar to a conventional mineral filter regardless of thickness. As the soil bridging is increased by the different particle sizes, a natural filter can be formed, thus reducing the possibility of clogging and any subsequent increase in hydrostatic pressure.

In an effort to apply functional design to a geotextile, filter criteria were created based on laboratory tests and in a few cases on practical experience. These filter criteria are based on the effective opening size (EOS) of the geotextile. Opening size for woven fabrics is relatively easy to determine; however, nonwoven, mechanically needle-punched fabrics are more difficult, requiring indirect measurements. Dry sieving as well as wet sieving (which better approximates changing flow natural conditions) have been used with marginal success, and deviations in results are frequent.

A common method of assessing filter criteria is the use of permeameter cells, which use the soils in question placed against a geotextile.

The permeameter is then subjected to a certain hydraulic gradient, and the flow and passage of fines is observed. A certain limit is set as to quantity of soil passage, which forms the basis for the criteria. A certain ratio between the grain size of the test soil and the opening size of the geotextile is then determined based on repeated tests. When comparing the available filter criteria developed by various researchers [*1–5*], it is apparent that they do not agree in their results, as shown in Fig. 3. For static hydraulic loads, high safety factors are needed when using some criteria, and it appears justifiable to base the measurements for needle-punched nonwovens on high proportional values, as has been confirmed by practical experience and research investigations. With increasing nonuniformity, the proportional value rises and grain sizes of larger fractions (which become larger and larger) are considered. The fines, however, become smaller and risk of erosion increases. In addition, suffusion (washing out of fines with stable grain structure) risk increases. Some filter criteria account for this problem based on filter requirements established by Cistin [*6*]; however, it must be realized that a maximum proportional limit value can not be exceeded.

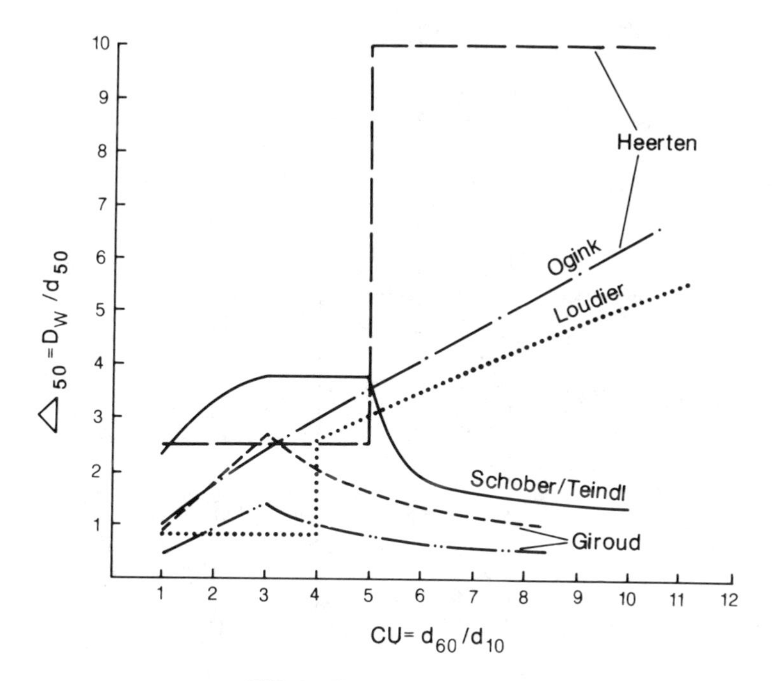

FIG. 3—*Existing design criteria.*

Filter Design Criteria for Needle-punched Nonwovens

Based on laboratory tests and years of practical experience, Chemie Linz has developed a series of empirical filter criteria for the Polyfelt needle-punched nonwoven fabrics as shown in Fig. 4. Teindl [*1*] found that the thickness of the geotextile influences its filter behavior (see Fig. 4). Differences between the different grades of Polyfelt can be seen on the design diagram, although their effective opening size (D_w) is of the same magnitude. It can also be seen that soil with a d_{50} smaller than 0.075 mm (0.003 in.) is critical. To use the design chart, the uniformity coefficient (CU) and the grain size d_{50} of the soil have to be known.

With dynamic loads, all particles of cohesionless soil acted upon by dynamic water movement which are smaller than the opening width of the geotextile are washed out. Therefore we need a filter that will safely retain at least part of the soil to be protected so that it is able to build up a "natural filter."

According to the various criteria available in the literature, the size of this soil fraction differs widely and probably depends directly on the project. It has to be taken into consideration that the soil particles which are larger in size than the effective opening size of the geotextile reduce

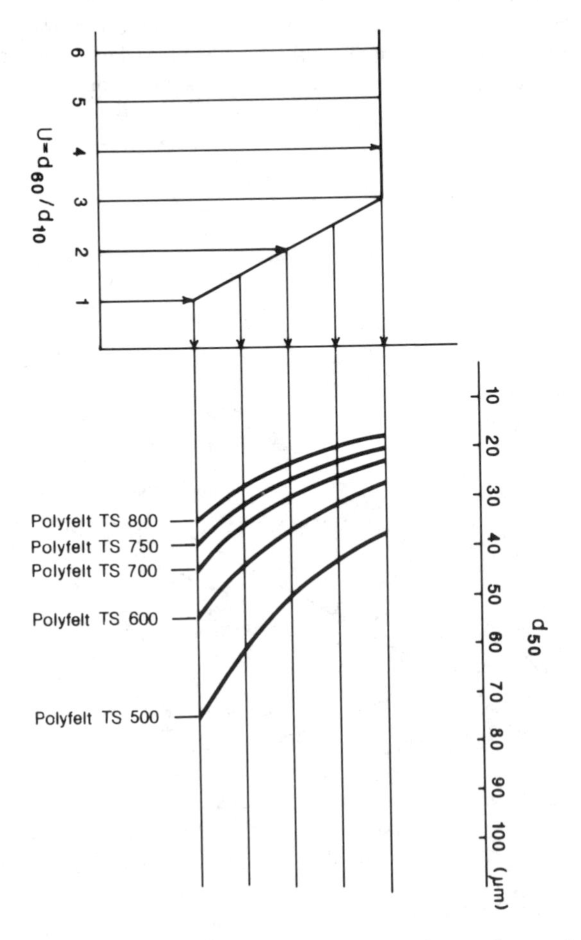

FIG. 4—*Filter design criteria for laminar, monodirectional water flow, and granular soil.*

the effective water energy. Fine-grain soil particles lying deeper are not affected, and the natural filter is stabilized. Its thickness and, therefore, the quantity washed out depends on soil composition and dynamic water energy.

With increasing nonuniformity of the soil, an increase in the ratio of particles smaller than the opening size of a geotextile may occur, increasing the risk of erosion through the geotextile. This can be designed for in a similar way as for static loads. Figure 5 represents a turbulent flow design chart based on data for Polyfelt TS fabrics. To use the turbulent flow design chart, the CU and the grain size d_{80} of the soil must be known. With CU known, the highest possible ratio D_w/d_{80} can be found. Multiplying d_{80} by the F ratio number found on the chart gives the required opening size (D_w) for the needle-punched nonwoven fabric.

With most criteria for mineral filters, a ratio d_{15} filter versus d_{80} soil (for example, Terzaghi $d_{15f}/d_{15s} = 4$) is used for guaranteeing sufficient permeability. Another possibility consists in comparing the permeability value k, a method which is used frequently with geotextiles. From the literature on existing criteria we can see that requirements differ widely.

Originally, calculations were based on the criteria for mineral filters. According to Terzaghi's design consideration, the k-value of the mineral filter must be 25 times that of the soil ($C_m = k_f/k_s$ 25). This value has been applied to geotextiles without considering the essentially different thicknesses of the geotextile and mineral filter. For static water flow conditions through the filter, a maximum hydraulic pressure dependent on the filter thickness and the k-value is required. This pressure must not increase substantially because of over design in thickness or an insufficient k-value of the filter (risk of hydraulic soil failure). Because the geotextile is relatively thin, the k-value may be correspondingly lower than that of a mineral filter for allowing the same water flow rate with the same hydraulic pressure as for the mineral filter.

The necessary thickness (t_F) of the mineral filter results from the fact that it has to correspond to 25 times the average filter grain diameter (d_{50}) to be stable geometrically. Out of the relation $C_m = k_f/k_s$ 25, a certain average grain diameter, and thus the layer thickness, results. On the basis of the ratio $k_f/t_f = k_g/d_g$, the k-value required for the geotextile may be determined. In

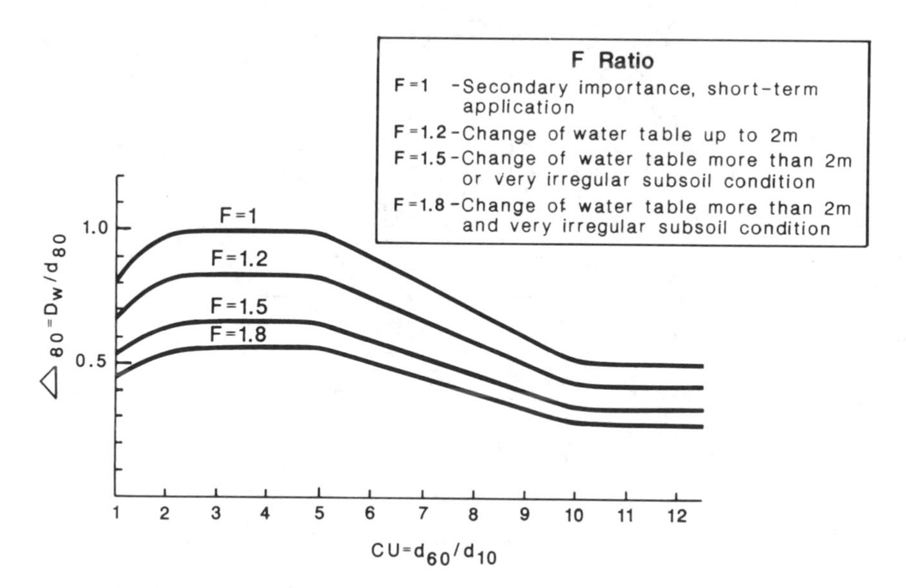

FIG. 5—*Filter design criteria for Polyfelt TS geotextiles under turbulent and alternating water flow when the uniformity coefficient (CU) and the granular soil grain size* (d_{80}) *are known.*

Fig. 6 the ratio C_p between the k-value of the subsoil (k_s) and the k-value of the geotextile (k_g) in accordance with the thickness of the mineral and geotextile filter is illustrated. Knowing the d_{50} grain size of the subsoil and the equivalent desired permeability ratio (C_p), the desired geotextile grade can be found.

There is very little information available about the filtration behavior of geotextiles and cohesive soils. With short fiber distances, clogging occurs more frequently, and, therefore, a minimum D_w is necessary. Haliburton and Wood [7] observed that with increasing silt content the clogging tendency rises. Woven tape-type fabric structures and thermally bonded nonwovens are especially affected.

In addition to the k-value, the fiber thickness also affects hydraulic filtration efficiency. Very fine fibers lie relatively close to each other and allow for bridge formation of very fine soil parti-

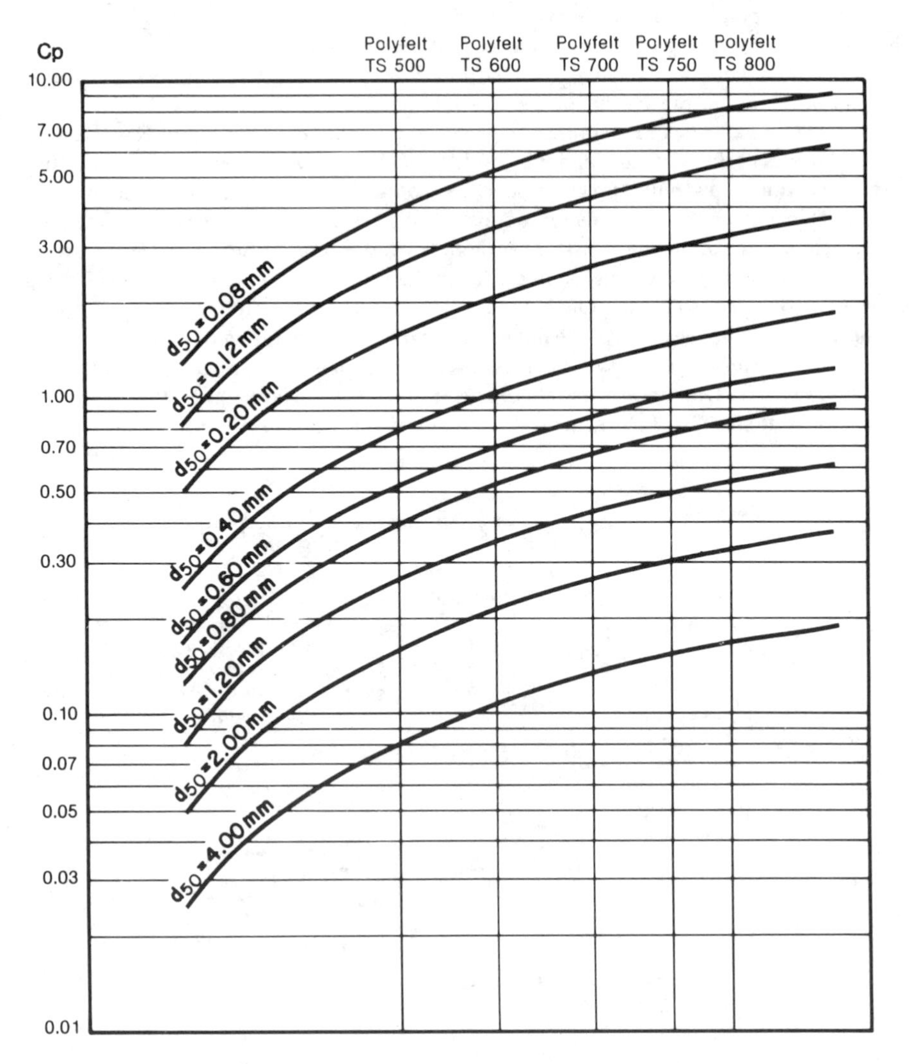

FIG. 6—*Graph for determining the desired equivalent permeability ratio* (C_p) *depending on the geotextile grade and the granular subsoil grain size* d_{50} *for* $k_g = C_p \cdot k_s$; *where* k_g = *geotextile permeability,* k_s = *permeability of soil, and* d_{50} = *soil particle size corresponding to 50% finer.*

cles. Sotton et al. [*8*] proved that fine fibers (approximately 3 deniers) mixed with thicker fibers influence the degree of clogging. This means that resulting fiber opening distances created with a combination of fine and coarse fiber composition may increase the risk of hydraulic pressure increase.

Up to now, based on practical experience, the opening width should be within the range of $D_w = 0.12$ mm (0.005 in.). With such a requirement, the retention property as well as hydraulic filter stability of mechanically bonded nonwovens is sufficient. The degree of soil compaction also affects the filter criteria. With increasing compaction, soil passage is reduced due to the higher stability of the grain structure [*9*]. This is another advantage that the mechanical needle-punched fabrics offer in that they adapt readily to uneven soil conditions and, therefore, secure soil position, whereas, with rigid products, cavities and consequently soil particle loosening may occur.

Construction Considerations

Site conditions and specific geotextile applications will generally govern the requirements for placement and construction using fabrics. Care must be taken during storage and placement to avoid damage and contamination to the fabric. When the geotextile is acting as a filter, any damage (that is, tears or holes) will greatly affect the filtering characteristics and possibly cause washout of supporting subgrade fines and subsequent localized failure.

In general, steps during construction should consist of the following:

1. Remove all large debris, stones, tree stumps, limbs, etc. and grade the area to effect a fairly level graded surface.
2. Fill any large depressions or holes in the slope and consolidate the fill if needed to prevent excessive settlement.
3. When placing the fabric, care should be taken to allow for slack. The geotextile should not be placed in a taught condition but should conform loosely with the subgrade without large irregular wrinkles or folds.
4. Place the geotextile with the longitudinal (machine) dimension in the direction of water flow, ensuring that any overlaps are in the direction of water flow.
5. If seams are required (that is, on the slope), the seam tensile strength should be equivalent to the fabric tensile strength criteria.
6. The placement of the protective armor must be in such a way as to prevent any puncturing or tearing of the geotextile. The type of protective cover material (that is, sandbags, stone, concrete block, etc.) will govern the method of placement to prevent damage to the fabric. Placement of cover material on slopes should always be from the base of the slope progressing up the slope and preferably beginning at the center of a geotextile covered section.

When placing stones, care must be taken not to allow stones weighing more than 34 kg (75 lb) to roll down a slope covered with fabric. Drop heights on fabric without a cushion layer must be rigidly controlled to prevent fabric damage and should be limited to 0.45 m (1.0 ft) for stones weighing less than 113 kg (250 lb). If larger armor stones are used, they must be placed carefully without dropping, as shown in Fig. 7. For installation with a cushion layer over the geotextile, stones up to 113 kg (250 lb) can be dropped from a height of 0.91 m (3 ft). However, larger stones must still be placed without dropping to prevent much disturbance of the cushion layer or damage to the fabric.

7. Slopes for erosion control applications should not be designed greater than 2.5 to 1 unless benching is utilized or the toe of the slope is stabilized and locked to prevent slippage. In any event, slippage between the soil slope/fabric or fabric/cover material must be minimized ideally. The soil-fabric friction angle should be determined for both the soil slope and the select

FIG. 7—*Placement of large armor stones.*

cover material or cushion material, and consideration should be given to saturated hydraulic conditions as well.

References

[1] Teindl, H., "Filterkriterien von Geotextilien," Report No. 153, Bundesministerium für Bauten und Technik, Hannover, Germany, 1980.

[2] Heerten, G., "Geotextilien im Wasserbau-Prufung, Anwendung, Bewahrung," Report No. 52, Mitteilungen des Franzius-Institutes für Wasserbau und Kusteningenieurwesens der Universität Hannover, Hannover, Germany, 1981.

[3] Ogink, H. J. M., "Investigations in the Hydraulic Characteristics of Synthetic Fabrics," Publication No. 146, Delft Hydraulics Laboratory, Delft, The Netherlands (Holland), May 1975.

[4] Giroud, J. P., "Filter Criteria for Geotextiles," *Proceedings, Second International Conference on Geotextiles,* Vol. 1, held in Las Vegas, NV, Industrial Fabrics Association International, St. Paul, MN, 1982.

[5] Loudiere, D. and Fayoux, D., "Filtration and Drainage with Geotextiles—Tests and Requirements," *Proceedings, Second International Conference on Geotextiles,* Vol. 1, held in Las Vegas, NV, Industrial Fabrics Association International, St. Paul, MN, 1982.

[6] Cistin, J., "Principles and Methods for the Design of Fabric Filters in Water Structures," *Vodni Hospodarstvi,* Vol. 30, No. 1, 1980, pp. 11-15.

[7] Haliburton, T. A. and Wood, P. D., "Evaluation of the U.S. Army Corps of Engineer Gradient Ratio Test for Geotextiles," *Proceedings, Second International Conference on Geotextiles,* Vol. 1, held in Las Vegas, NV, Industrial Fabrics Association International, St. Paul, MN, 1982.

[8] Sotton, M., Leclerq, B., Fedoroff, N., Fayouz, D., and Paute, J. L., "Contribution & l'Etude du Colmatage des Geotextiles. Approch Morphologique," in *Proceedings, Second International Conference on Geotextiles,* Vol. 1, held in Las Vegas, NV, Industrial Fabrics Association International, St. Paul, MN, 1982.

[9] Wittmann, L., "Soil Filtration Phenomena of Geotextiles," *Proceedings, Second International Conference on Geotextiles,* Vol. 1, held in Las Vegas, NV, Industrial Fabrics Association International, St. Paul, MN, 1982.

Soil Reinforcement

Bernard Myles[1]

A Review of Existing Geotextile Tension Testing Methods

REFERENCE: Myles, B., **"A Review of Existing Geotextile Tension Testing Methods,"** *Geotextile Testing and the Design Engineer, ASTM STP 952,* J. E. Fluet, Jr., Ed., American Society for Testing and Materials, Philadelphia, 1987, pp. 57–68.

ABSTRACT: The review assesses the existing geotextile tension testing procedures in both North America and Europe with particular reference to the difference between the conditions of local and total stress imposition.

In local stress testing, the adaption of the California Bearing Ratio (CBR) soil test is described as a complement to the more commonly used Grab Test. The importance of the failure mode and fabric structure integrity under stress are also highlighted.

Attention is drawn in the unidirectional (total stress) test to the relevance of sample width, variation of strain rate, and the different definitions of modulus. Also, the requirement to relate tensile properties to a time base and a means of presenting these properties is shown.

KEY WORDS: geotextile, tension testing

The aim of this paper is to review and discuss the differing tension testing methods that are applied to geotextiles in Europe and North America. The author would like to recognize the considerable efforts of the fellow members of ASTM Committee D-13 on Textiles and subsequently Committee D-35 on Geotextiles, Geomembranes, and Related Products in developing the 200-mm wide width strip method and recommends that this paper be read in conjunction with ASTM Test Method for Tensile Properties of Geotextiles by the Wide Width Strip Method (D 4595-86) and ASTM Test Method for Breaking Load and Elongation of Textiles (Grab Method) (D 4632-86).

Introduction and Objectives

A geotextile tension test procedure must be defined and performed in such a way as to closely resemble the anticipated service conditions. At the same time the test should be sufficiently simple to enable comparative and control procedures.

Test data requirements will vary from empirical index to sophisticated design parameters. The concept of local and total load application has implications for the testing procedure, that is, the uniform tension induced in a fabric by embankment settlement will vary greatly from the localized stresses induced by a wheel load acting perpendicular to the plane of the fabric (Fig. 1). While the desire to have a single geotextile test for each specific property is widespread, it should be recognized that the variations in field conditions may give rise to forces within the geotextile which call for more than one test in order to evaluate the ability of the geotextile to withstand these diverse stresses.

[1]Consulting engineer, Catbrook, Gwent, United Kingdom NP6 6NA, formerly technical manager, ICI Fibres Geotextile Group, Pontypool, United Kingdom NP4 8YD.

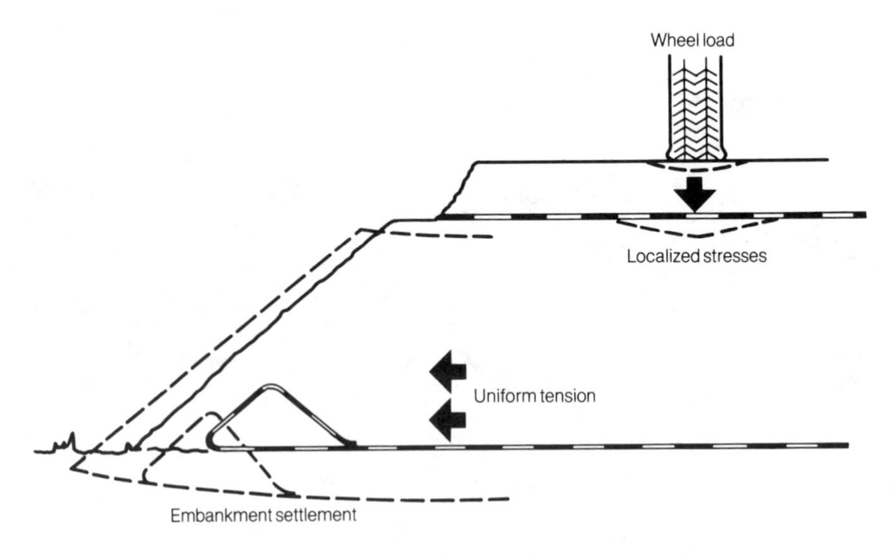

FIG. 1—*Geotextile tensioning by local or total load.*

Test Criteria

Simplicity of procedure and clarity of concept are essential if a geotextile tension test is to gain wide acceptance and use throughout the industry, particularly by manufacturers. Rates of strain in common use in existing "textile" tension tests are considered to be too high for geotextiles; a lower rate would be more realistic. However, by comparison with geotechnical rates of strain, geotextiles have a very different threshold of sensitivity (Fig. 2).

A choice of loading conditions between constant rate of strain and constant rate of load have historically been determined more by the availability of testing equipment considerations than by philosophical argument. However, constant rate of strain procedures have predominated in both textile and soil testing, and this would indicate its eventual universal adoption for geotextiles.

Size of Specimen

The size of the specimen to be tested is of extreme importance. Geotextiles are not normally tensioned "in isolation" for separation or reinforcing applications but are nearly always encapsulated in soil, thus restraining lateral deformation. The choice of specimen size and aspect ratio must take account of this condition. Geotextiles are produced and used in large format, rolls of 4.0 to 5.0 m wide being the norm. Therefore, the width of the specimen tested must adequately represent these dimensional considerations (Fig. 3). A comparison of adequate width and appropriate aspect ratio, when applied to geotextiles of high strength, will place significant demands on machine capacity, fabric fixing, and on the gripping mechanism (Fig. 4). Figure 5 shows a 1000-mm wide width test being used to assess aspect ratio.

Existing Geotextile Tension Tests

The existing geotextile tests (Fig. 6) can be defined in two main categories:

1. Application of a load perpendicular to the geotextile.
2. Application of a load in the plane of the geotextile.

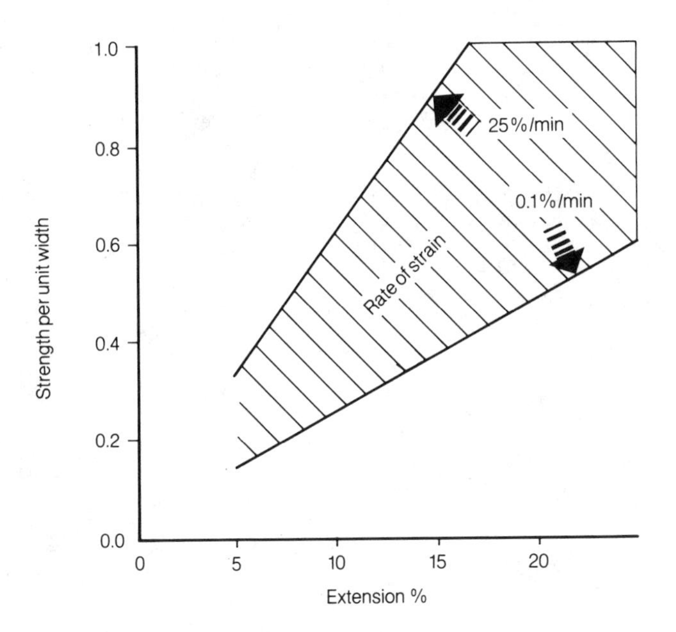

FIG. 2—*Load/extension envelope for differing rates of strain.*

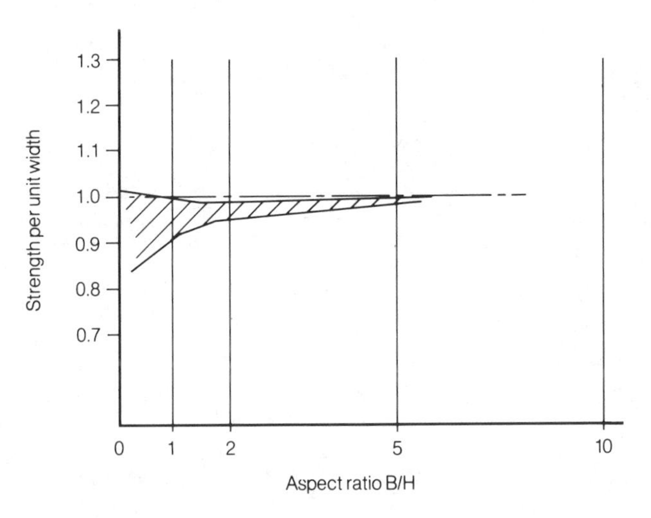

FIG. 3—*Influence of sample width on strength-to-height ratio.*

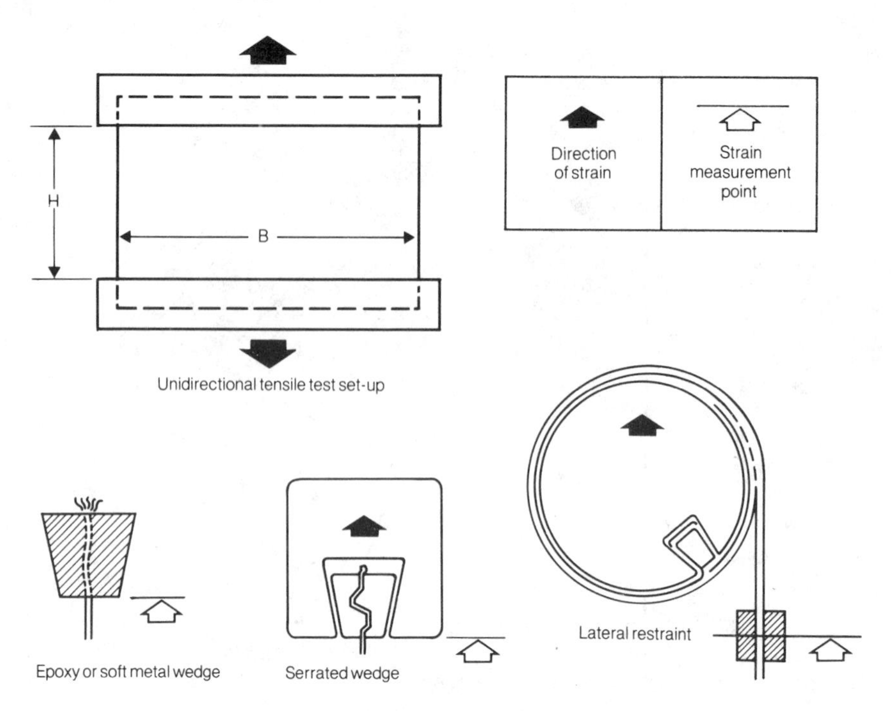

FIG. 4—*Unidirectional test setup. Geotextile clamping systems.*

California Bearing Ratio (C.B.R.)

A diaphragm of geotextile is clamped in a 150-mm (6-in.) CBR mould, and a 50-mm (2-in.) plunger is applied centrally. The distension and push through-force are recorded.

Cylindrical Sleeve Test (Saint-Brieuc)

The geotextile is seamed into a 100-mm-diameter cylinder. The ends are fixed 200 mm apart in circular clamps, and an internal pressure is applied by a flexible membrane. The extension at the circumference and rupture pressure are noted.

Circular or Rectangular Burst Test

The fabric is clamped onto a flexible membrane, the specimen size varying between 36 mm diameter and 200 by 800 mm rectangular. The fabric is distended and the rupture pressure noted.

Crucifix Biaxial Test

The four sides of a rectangular or crucifix form of geotextile is clamped on all sides and force applied in cross and length direction simultaneously. The loads and extensions are recorded.

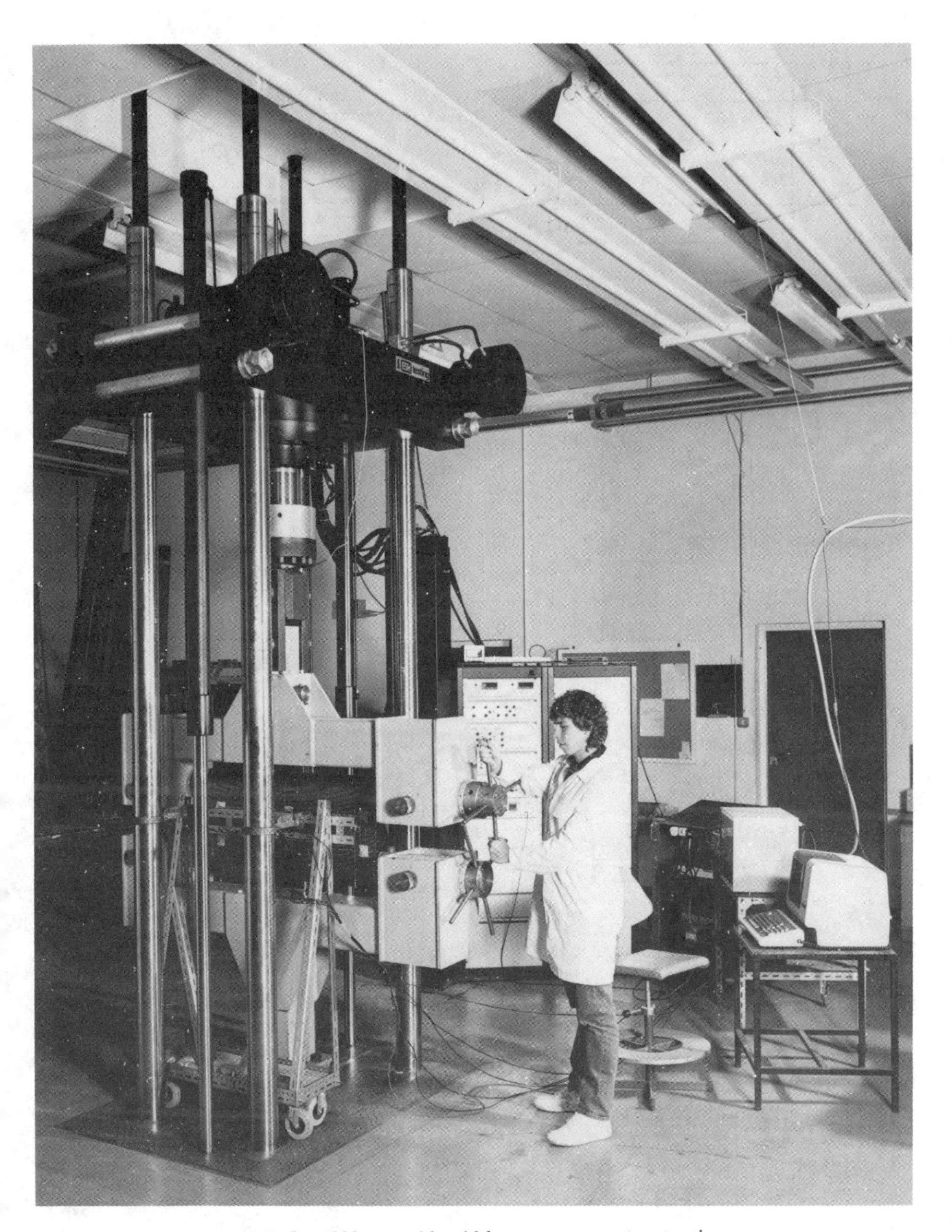

FIG. 5—*1000-mm wide width test to assess aspect ratio.*

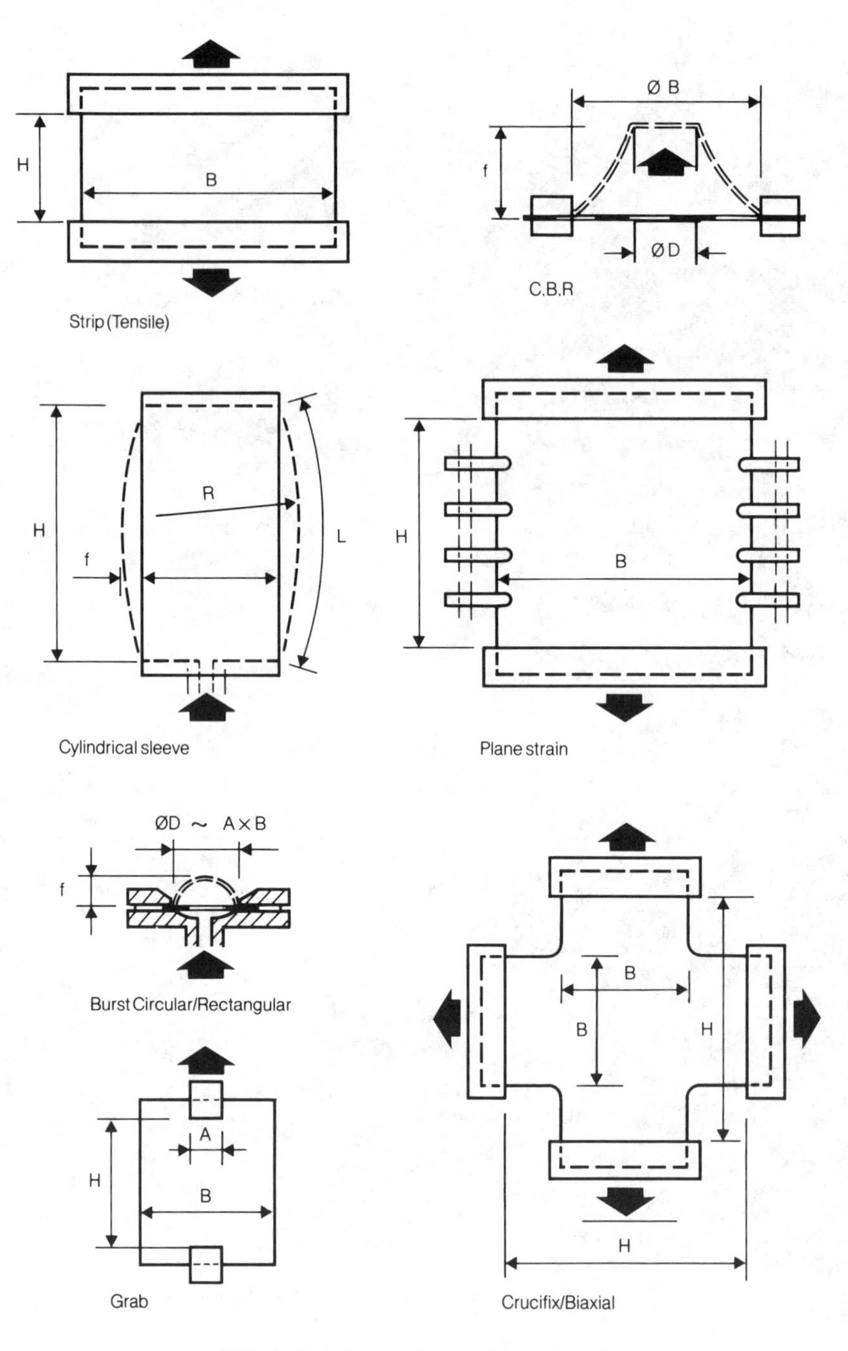

FIG. 6—*Existing tension test for geotextiles.*

Grab Test

A localized force 25 mm wide is applied across the opposite sides of a rectangular fabric (100 by 100 mm). Load and extension are recorded.

Strip Tension Test

Widths of geotextiles, varying from 50 to 500 mm with aspect ratio varying from 1:4 to 10:1, are tensioned uniaxially and uniformly across the width. The breaking load and breaking extension are recorded.

Laterally Restrained Tension Test

Widths of geotextile of 200 or 100 mm with aspect ratios of 1:1 to 2:1 are tensioned uniaxially and uniformly across the width. The fabric is restrained laterally from deformation by mechanical constraint or granular encapsulation. The breaking load and breaking extension are recorded.

Choice of Geotextile Test

The wide variation in the current test procedures would seem to indicate a great diversity of philosophies or parameters and property requirements. This is not the case in practice. Although the procedures may vary, the objectives are similar. What has happened is that a variety of tension testing procedures originating from geotextile manufacturers has now been complemented by modified compression testing procedures from the geotechnical area. The development of applications, dominated as it was initially by separation, gave rise to an emphasis on local load and deformation testing, for example, CBR, grab, burst; while at the same time the original narrow width tension testing procedures were modified to give more representative geotextile information. There has consequently emerged two types of tests:

1. A localized loading test giving a good empirical indication as to the robustness of the geotextile and its ability to function as a separator.

   (*a*) The adoption of the CBR soil test procedure has gained widespread acceptance, particularly in Europe, primarily due to the availability of the testing equipment and the simplicity of the testing procedure, not least the ease with which the fabric can be clamped. However, this method has not yet gained wide acceptance within the United States. Figure 7 shows a CBR test using a tension testing machine.

   (*b*) The grab tension test has a long history within the textile industry and can be found in most international tension testing procedures, and while this testing procedure has lost favor and acceptance in Europe, it still remains widely used and accepted in the United States.

2. A unidirectional loading test giving an indication of the stress/strain characteristic and the ability to develop a tensile load.

   (*a*) The unidirectional test is exemplified by the ASTM 200-mm wide width strip tension test (D 4595-86). This was developed from the conventional 50-mm strip tension textile test, a test which is still widely used for geotextiles, primarily in quality control and indexing. The 200-mm test also owes its existence in part to the French 500-mm wide width tension test.

   (*b*) The present degree of international acceptance of the 200-mm-wide test has been a long and arduous process, not just because of the wide variations in geotextiles to be tested, but because the procedures have necessitated constant adaptation and adjustment of equipment due to the progressive development of high-strength geotextiles and the continuing re-

FIG. 7—*CBR test using tension testing machine.*

quirement to develop adequate clamping mechanisms and suitable methods of measuring the extension. Figure 8 shows a 200-mm wide width test using a roller grip.

Presentation of Results

All geotextile testing results should be presented in the form of a mean and standard deviation. The results can be presented in the form of a maximum load or breaking load with appropriate extensions, or they can be presented in graph form which shows in more detail the stress/strain behavior. In addition to the breaking load and breaking extension, or maximum load and

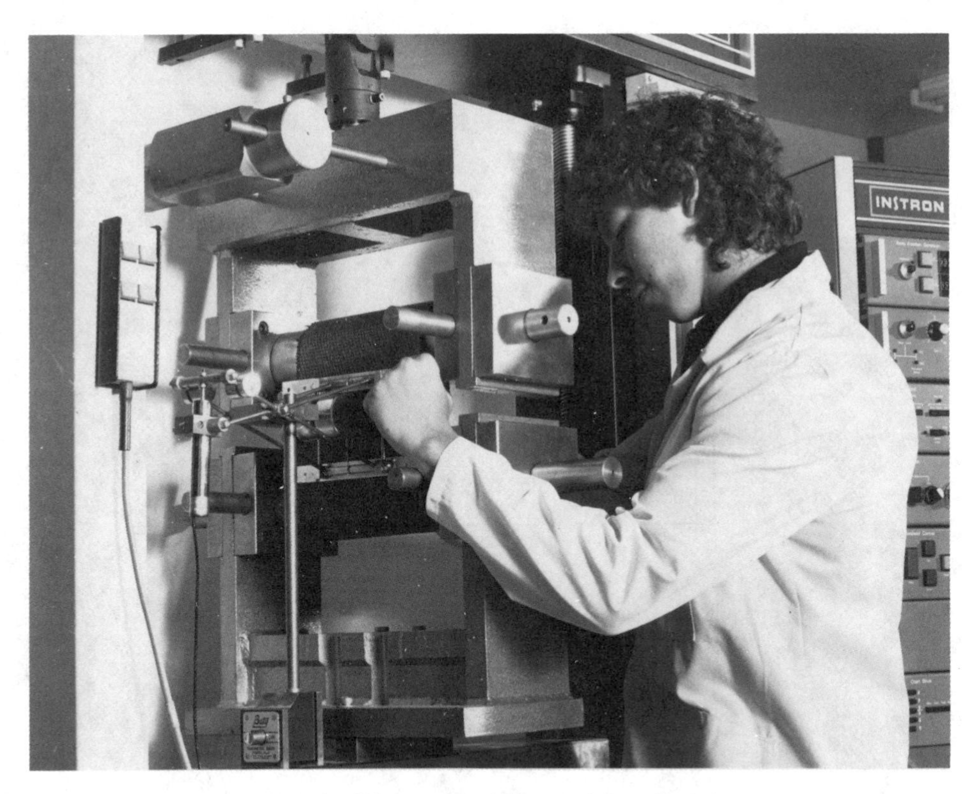

FIG. 8—*200-mm wide width test using a roller grip.*

maximum extension, an appropriate modulus should be defined. To be meaningful, the elastic response function (modulus) must relate to the predicted working stress/strain range of the geotextile which, in turn, must take into account the possible and acceptable soil strain.

While the requirement for reporting modulus is generally nonmandatory, when geotextiles are to be used in a reinforcing application, the importance of the "working modulus" cannot be overstated. High-strength and reinforcing geotextiles often have a Hookean region. Therefore the offset modulus as described in the ASTM wide width strip method (D 4595) is well suited to describe the "working modulus." In addition, the extension intercept should also be reported (Fig. 9).

The initial and tangential modulus are often unsuitable as a definition of "working modulus," unless an exact condition of strain can be assured. In the United States the secant modulus has become widely used, often defined by the 0.05 or 0.1 strain condition. However, Fig. 10 demonstrates the difficulty in comparing two geotextiles of differing elastic response by means of a single strain secant modulus value. If the elastic response is to be defined by secant, then the working strain range should be bracketed by two strain secant values.

Interpretation of Tension Testing Results

Localized Load (CBR-Grab)

An estimation of maximum and minimum field deformation perpendicular to the geotextiles should be made and compared with the load/distension development. The type of failure

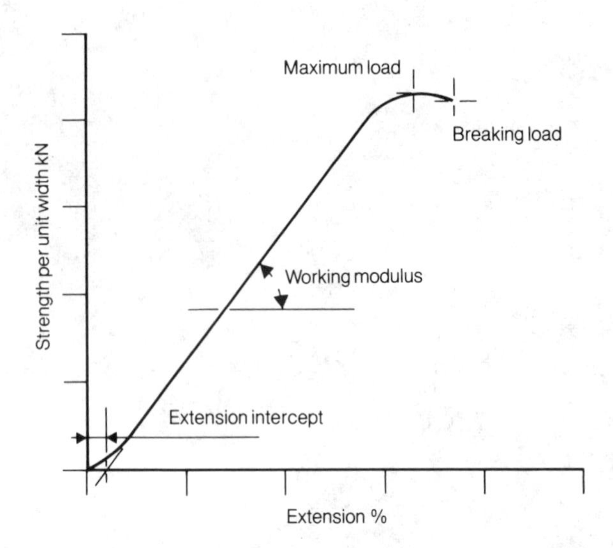

FIG. 9—*Unidirectional load by wide width load/extension curve.*

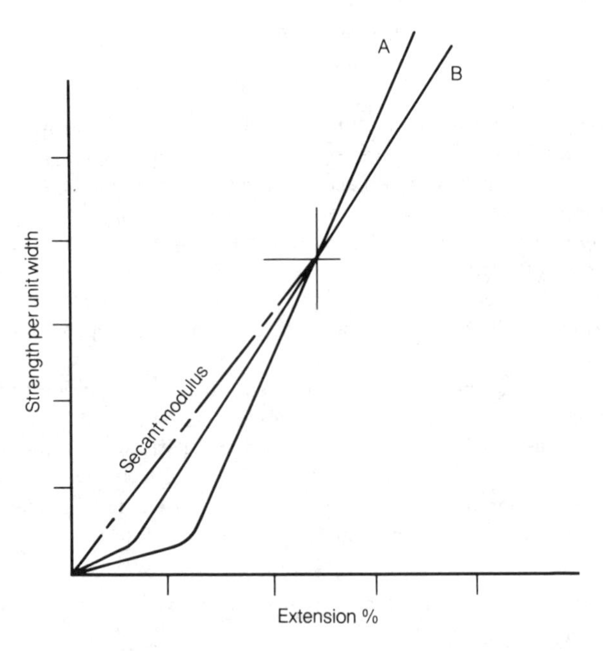

FIG. 10—*Differing geotextiles A and B with the same secant modulus.*

should be noted, that is, brittle or soft (Fig. 11). The stability of the fabric geometry should also be noted, particularly if the geotextile has to perform as a separator/filter on fine-grain soils.

Unidirectional Load (Wide Width Strip Method)

The breaking load and breaking extension should be used together with a factor of safety to establish the working condition.

The anticipated deformation to achieve the working load should be established, not only by means of the final structural configuration but also with reference to the mode of and stages of construction.

The "working modulus" and extension intercept will be modified by soil encapsulation. This is not particularly significant in oriented fabric structures but could be pronounced in highly extendable geotextiles.

Additional Information to Support Tension Testing Results

Localized Load

Deformation of the fabric structure may be pronounced in the zone of high stress produced by localized load. This is particularly important if this brings about significant changes in the filter characteristics of the geotextile. In addition to visual inspection, it may be advisable to assess pore size change by means of dry sieving, airflow, or water flow.

Unidirectional Load

The prediction of creep and stress relaxation at the anticipated working loads is of extreme importance if the geotextile is to sustain its load over a set period. This information can be combined with a short-term stress/strain plot in a "BIM" diagram (Fig. 12).

In addition, the soil/geotextile friction characteristic should be defined as this is the principal means of transferring a tensile load from the geotextile into the soil.

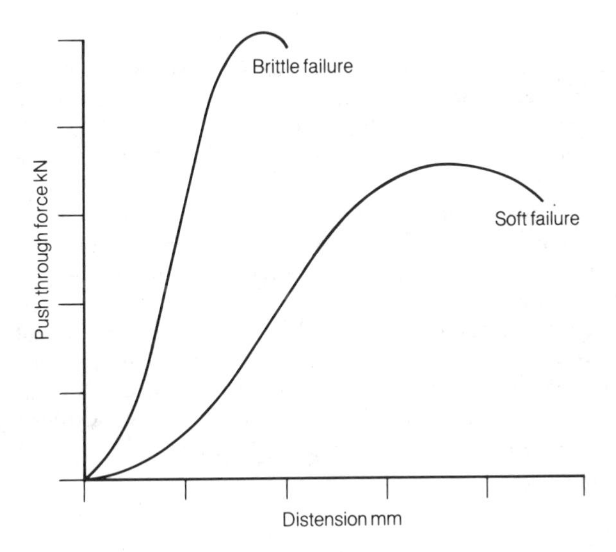

FIG. 11—*Localized load by CBR plunger load/distension curves.*

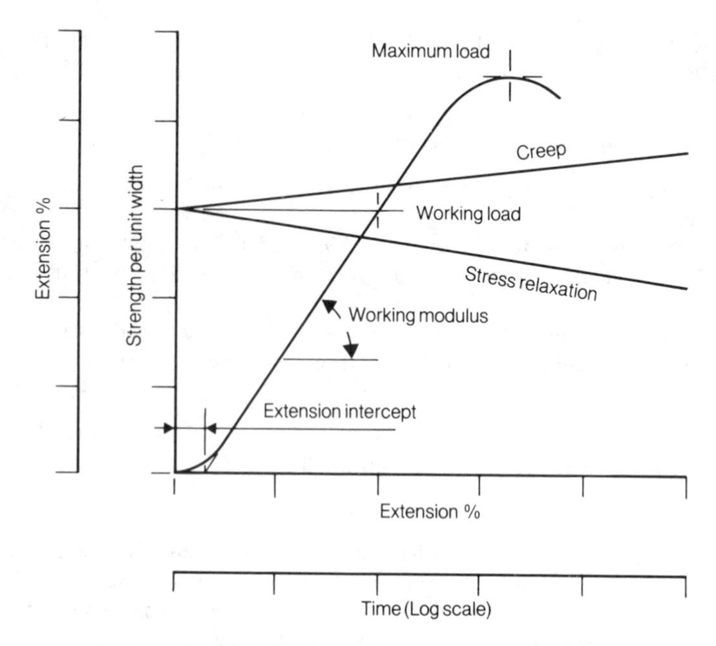

FIG. 12—*Combined load-extension-time curve (BIM diagram).*

Jointing and Seaming

Methods of jointing and seaming are an integral part of the tensile properties of geotextiles, as the lateral strength in a geotextile is often dependent on the ability of the joint or seam to transfer load.

Conclusions

While the concept of tension testing geotextiles is closely linked to the use of geotextiles as an element in soil reinforcement, it must be remembered that the use of geotextiles as subgrade/subbase separators still dominates, and the perception of "strength" must be broader than simply as a stress/strain diagram. The problems of puncture, tear, and abrasion must also be taken into account.

However, all things considered, there now exist adequate methods for tension testing geotextiles. Few people will claim these tests are perfect, but much effort and time has been expended to develop reproducible and usable results to give engineers confidence to integrate the structural properties of geotextiles into their designs.

Not only must the manufacturing section of the geotextile industry accept and use these proven test methods, but specifiers, authorities, and geotechnical designers must insist on geotextile tensile information being presented in a uniform, understandable, and comparative form. Only in this way will better use be made of geotextile properties, thus leading to more cost-effective geotechnical solutions.

Rudolph Bonaparte,[1] *Robert D. Holtz,*[2] *and J. P. Giroud*[3]

Soil Reinforcement Design Using Geotextiles and Geogrids

REFERENCE: Bonaparte, R., Holtz, R. D., and Giroud, J. P., **"Soil Reinforcement Design Using Geotextiles and Geogrids,"** *Geotextile Testing and the Design Engineer, ASTM STP 952,* J. E. Fluet, Jr., Ed., American Society for Testing and Materials, Philadelphia, 1987, pp. 69–116.

ABSTRACT: This paper provides an overview of the design of soil structures reinforced with polymeric materials. Applications for soil reinforcement are reviewed and categorized according to the type of soil structure, type of load, and function and location of reinforcement. Approaches to design are reviewed, and it is shown that most current design procedures are based on simple extensions of classical limit equilibrium analyses.

Properties of polymeric reinforcement relevant to design are presented. They are subdivided into tensile properties and soil-reinforcement interaction properties. Use of wide-strip, constant-load creep tests are recommended for evaluation of tensile properties for design of permanent soil structures. Soil-reinforcement interaction characteristics are evaluated in terms of direct sliding of soil over reinforcement and pullout of reinforcement from soil. Direct sliding characteristics should be evaluated using the direct shear test. Pullout characteristics for geotextiles can be approximated from direct shear test results; however, pullout characteristics for geogrids must be evaluated using pullout tests.

The paper concludes with a detailed presentation of simple limit equilibrium design analyses for three of the most commonly encountered types of reinforced soil structures: reinforced slopes, reinforced soil walls, and reinforced embankments over weak foundations. Specific recommendations are made on determination of reinforcement properties for design of these three types of structures.

KEY WORDS: soil reinforcement, geotextiles, geogrids, wide-strip tension test, creep test, direct shear test, pullout test, slope stability, retaining walls, embankments

The design engineer dealing with soils has two choices: use the soil as it is or improve it. Ever since geotechnics appeared as a separate engineering discipline more than 50 years ago, design engineers have taken the soil as it is, with all its deficiencies and weaknesses, as compared to other common engineering materials such as wood, steel, and concrete. Similar to these other materials, soil is strong in compression. However, unlike them, soil has virtually no tensile strength. Soil reinforcement using high tensile strength inclusions can give a soil mass tensile strength.

Soil reinforcement is not a new concept. Early examples of soil reinforcement include the ancient ziggurats found in Iraq, which are more than 3000 years old. Reed-reinforced earth levees were constructed along the Tiber River by the Romans. The modern uses of soil reinforcement appeared in the 1960s with the development of Reinforced Earth retaining walls and geotextile stabilization of haul roads and access roads. To date, thousands of retaining struc-

[1]GeoServices Inc. Consulting Engineers, Boynton Beach, FL 33435 (formerly, technical director, The Tensar Corp., Morrow, GA 30319).

[2]Professor, Soil Mechanics Laboratory, Purdue University, West Lafayette, IN 47907.

[3]GeoServices Inc. Consulting Engineers, Boynton Beach, FL 33435.

tures have been constructed with the Reinforced Earth technique and many thousands of roads with geotextiles.

This paper presents an introduction to soil reinforcement using polymer reinforcement materials such as geotextiles and geogrids. The paper covers applications, theory, and design. The approach taken in this paper is to:

1. Understand basic soil reinforcement principles, while recommending simple design methods.
2. Understand the complex behavior and limitations of geotextiles and geogrids, while recommending simple tests to generate "safe design values" for relevant material properties.

The paper will begin by introducing different types of reinforced soil structures. Then, the basic principles of soil reinforcement will be reviewed, and reinforcement materials will be presented. Testing procedures and interpretation of test results will then be discussed. Finally, simple design methods and selection of reinforcement properties for design will be presented for three of the most commonly encountered types of reinforced soil structures: reinforced slopes, reinforced soil walls, and reinforced embankments on weak foundations.

Types of Reinforced Soil Structures

The main types of reinforced soil structures are classified as indicated in Table 1. These structures can be subdivided into two broad categories, earth structures and load supporting structures:

1. *Earth structures* include slopes, walls, embankments, low-permeability soil layers used in dams and waste containment facilities, and some "soil layers on nonuniform foundations" such as roads and embankments over karst topography. Earth structures do not normally support significant external loads, and the primary design consideration is the stability of the structure under its own weight.
2. *Load supporting structures* include flexible pavements, unpaved roads, railroad track structures, and load supporting pads such as drilling pads, fabrication yards, and construction staging areas. These structures are usually stable under their own weight, and the primary design consideration is the structure's ability to support the applied loads (usually traffic loads) with limited deformation.

In these applications, the reinforcement is either located inside the structure (see "internal reinforcement" in Table 1) or at the interface between the structure and the foundation soil. A brief presentation of the reinforced soil structures listed in Table 1 follows.

Reinforced Slopes and Soil Walls

Construction—Reinforced slopes and soil walls are typically constructed with alternating horizontal layers of compacted soil and reinforcement. With reinforced soil walls and very steep reinforced slopes, a facing may be necessary to prevent localized surface erosion and sloughing along the exposed side of the reinforced soil mass. Many types of facings can be used. Two of the more common include concrete panels (segmental or full-height) and "wrap around" facings (that is, a facing provided by wrapping the reinforcement around the vertical face of the adjacent compacted soil layer).

Role of the Reinforcement—In these applications, the reinforcement strengthens the structure by adding tensile strength to the soil mass and by increasing soil strength as a result of increased soil confinement. This strengthening permits construction of stable soil structures at angles steeper than the soil's angle of repose and/or higher than would be possible without

TABLE 1—*Classification of soil reinforcement structures.*

Reinforcement Location	Type of Structure	Typical Applications	Load		Class
Internal	Reinforced slopes	Embankments Landslide repairs Excavations Dams, dikes, and levees	Essentially dead load	Essentially distributed load (gravity)	Earth structure
	Reinforced soil walls	Retaining walls Bridge abutments Interchange structures Embankments Barrier walls (for noise, boulders . . .) Blast walls and military shelters Dams, dikes, and levees			
	Sloped soil layers	Clay liners Earth covers			
Interface	Embankments on weak foundations	Embankments Dams, dikes, and levees			
Combined internal/interface	Soil layers on nonuniform foundations	First layer of earthwork construction Clay liners on nonuniform soils Landfill clay covers over landfill waste		Essentially concentrated load	
		Road structures on nonuniform soils	Dead and/or live load		Load supporting structure
	Load supporting pads	Construction equipment pads Heavy vehicle access Fabrication yards and staging areas Intermodal facilities			
Interface	Unpaved roads on weak foundations	Access roads Haul roads First layer of earthwork construction	Essentially live load (traffic load)		
Internal	Flexible pavement structures	Paved roads Industrial storage yards Parking lots Low-volume high performance unpaved roads			
	Ballast reinforcement	Railroad track structures			

reinforcement. For a given soil structure, the inclusion of tensile reinforcement will increase the factor of safety against failure and reduce soil movements.

Applications—These earth structures include:

1. Reinforced slopes (see Table 1) used for embankment construction (Fig. 1*a*), natural slope stabilization (Fig. 1*b*), and excavations (Fig. 1*c*). Typical applications include: highway embankment widening, embankment construction for highways, dikes, levees, etc. (Fig. 1*a*); landslide repair (Fig. 1*b*); silos or storage facilities for ore, coal, or aggregate (Fig. 1*c*).
2. Vertical or near-vertical reinforced soil walls (see Table 1) having either concrete facings (Fig. 2*a*, 2*c*, 2*d*, and 2*e*), "wrap-around" facings (Fig. 2*b* and 2*f*), or other facings such as timber, brick, or gabions. Typical applications include: retaining walls (Fig. 2*a* and 2*b*); bridge abutments (Fig. 2*c*); structures (Fig. 2*d*) used as dikes, jetties, quay walls, waterfront structures, architectural walls, military shelters, blast walls, noise barriers, boulder barriers, etc.; vertically faced structures used to raise existing dams (Fig. 2*e*) or construct new dams (Fig. 2*f*).

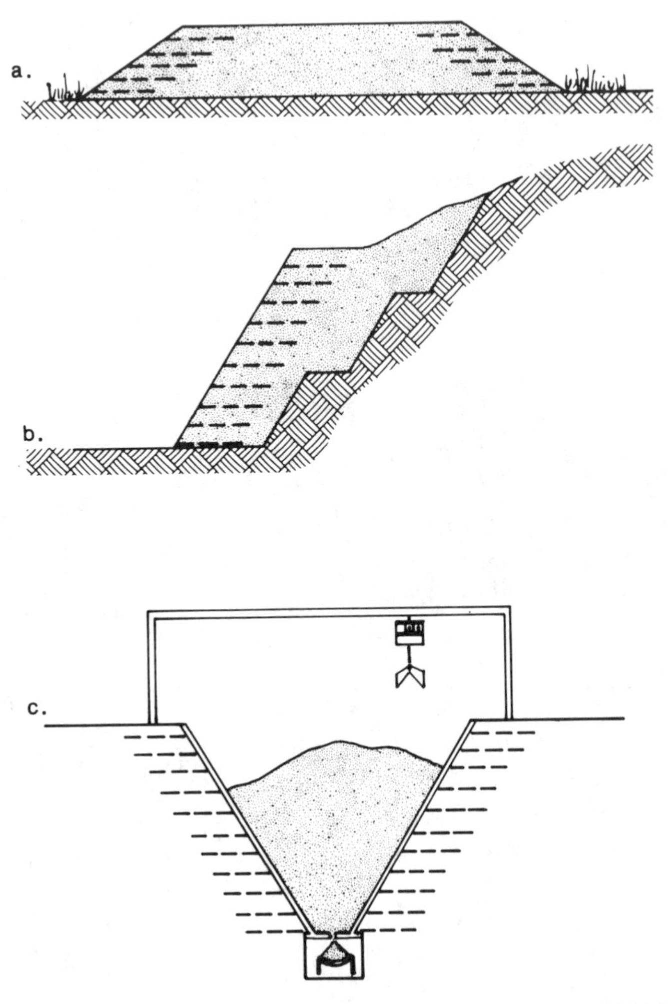

FIG. 1—*Reinforced slopes:* (a) *embankment;* (b) *landslide repair; and* (c) *silo.*

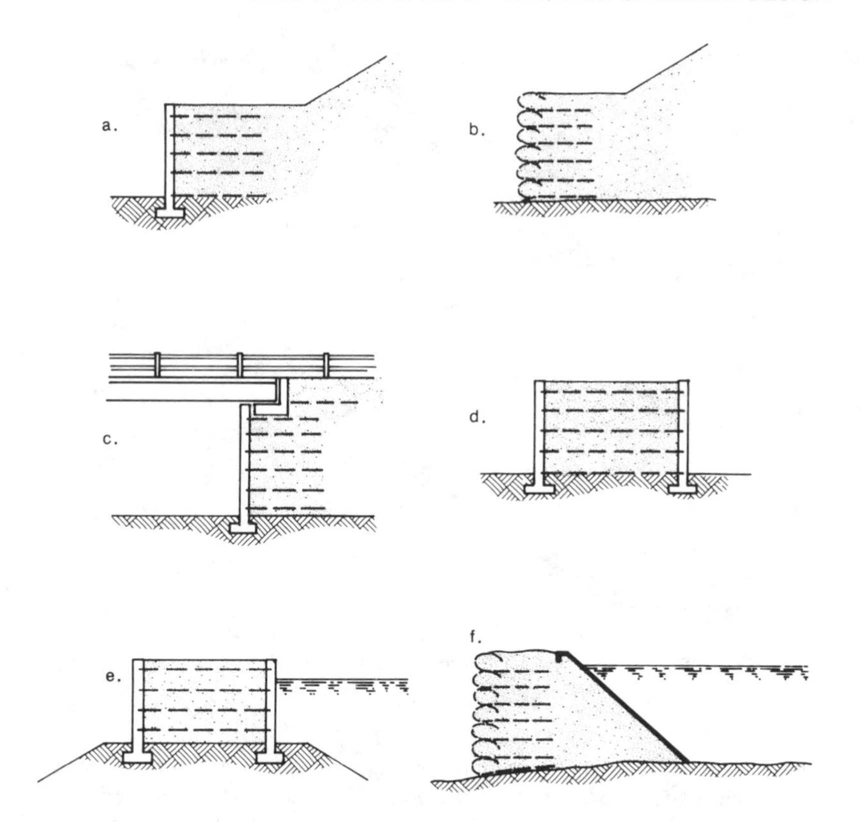

FIG. 2—*Reinforced soil walls:* (a) *concrete faced retaining wall;* (b) *"wrap-around" faced retaining wall;* (c) *bridge abutment,* (d) *dike or waterfront structure;* (e) *dike or dam raising; and* (f) *dam.*

Sloped Soil Layers

Construction—Sloped soil layers (such as clay liners or earth covers associated with geomembrane liners) have been constructed with reinforcement placed either within or at the bottom of the soil layer. The reinforcement is typically parallel to the slope surface and is anchored in a trench at the top of the slope.

Role of the Reinforcement—In these applications, the reinforcement reduces or prevents cracking and/or downslope movement of the soil layer due to insufficient frictional resistance along an underlying interface. As a result of the mobilized tension in the reinforcement, the downslope shear stresses along the underlying interface are reduced and a stable soil layer can be constructed on a slope steeper than would be possible without reinforcement.

Applications—Clay liners and earth covers on slopes (see Table 1) are typically used in solid waste disposal facilities such as landfills and in liquid surface impoundments such as industrial ponds and water storage reservoirs. Low permeability clay liners (Fig. 3*a*) are used as liquid barriers, and it is important that they do not crack. Earth covers (Fig. 3*b*) are used to protect geomembrane liners, and their stability is often impaired by the low friction coefficient between the compacted soil cover and the polymeric geomembrane. The use of reinforcement permits construction of clay liners and earth covers on slopes steeper than permitted without reinforcement, thereby increasing the usable volume of the landfill or impoundment.

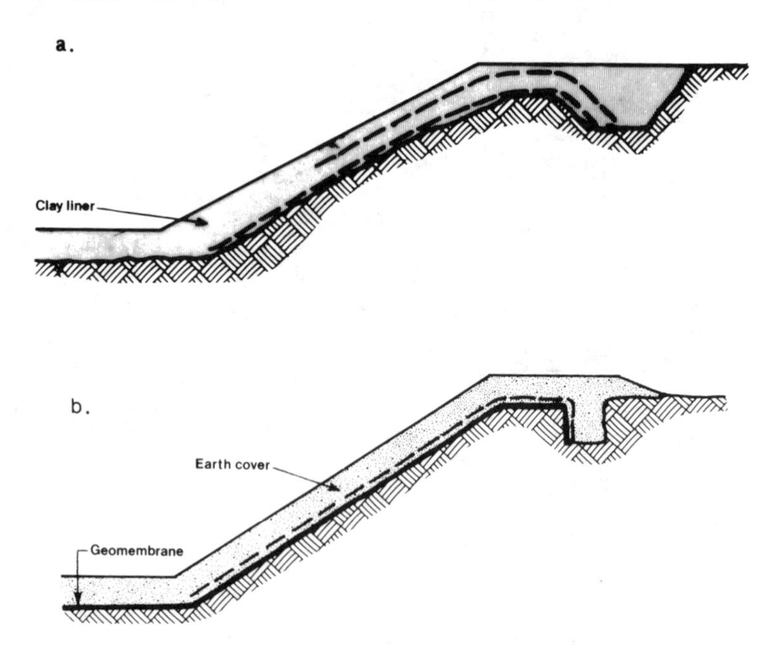

FIG. 3—*Sloped soil layers:* (a) *clay liner; and* (b) *earth cover on a polymeric liner.*

Reinforcement placed on a slope can act as a slip surface. Selection of reinforcement with sufficiently high soil-reinforcement interface friction is critical in this application.

Embankments on Weak Foundations

Construction—A layer of reinforcement is placed on the natural soil or on a thin working pad, and the embankment is constructed in the conventional manner on top of it. Sometimes, two or three layers of reinforcement are used at or near the base of the embankment.

Role of the Reinforcement—There are at least two categories of weak foundations. If construction is to take place on a site underlain with a uniformly weak soil deposit such as soft clay, peat, or muskeg (Fig. 4*a*), the role of the reinforcement is to increase the factor of safety against a slip-surface failure of the embankment and foundation soil, and to reduce lateral spreading and cracking of the embankment. (Note: Reinforcement will not significantly reduce embankment settlements associated with time-dependent consolidation of uniform foundation soils.) If the foundation soil is locally weak because of lenses of soft clay or peat, or because of sinkholes (Fig. 4*b*), the role of the reinforcement is to bridge the weak spots (by the tensioned membrane effect and by the promotion of soil arching) in order to reduce the risk of localized failure and/or reduce differential settlements.

Applications—These earth structures (see Table 1) include highway embankments, dikes, levees, dams, etc. (Fig. 4*a* and 4*b*). Embankments with reinforced slopes built on soft soils (Fig. 4*c*) incorporate a combination of the reinforcement functions presented in Fig. 1*a* and Fig. 4*a*.

Soil Layers on Nonuniform Foundations

Construction—A layer of reinforcement is placed under and/or in soil layers constructed on nonuniform foundations to provide localized support. Sometimes two or three layers of reinforcement are used at or near the base of the soil layer.

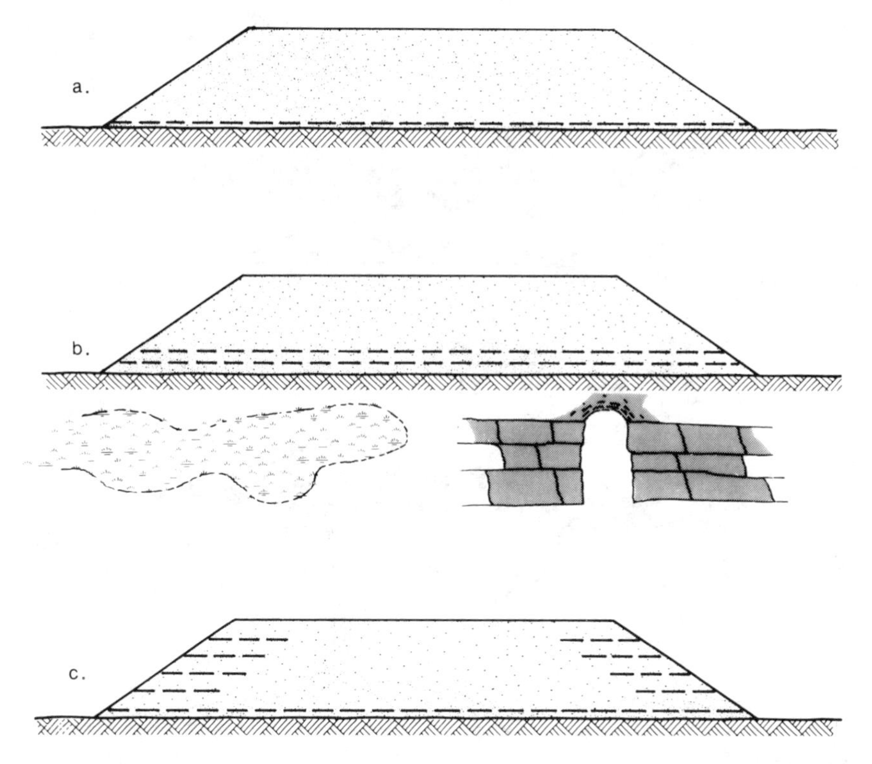

FIG. 4—*Embankments on weak foundations:* (a) *embankment on a uniform weak foundation;* (b) *embankment on a locally weak foundation with lenses of clay or peat, sinkholes, etc.; and* (c) *embankment with reinforced slopes, built on a uniform weak foundation.*

Role of the Reinforcement—The reinforcement is used to transfer stress away from zones of weakness, assist soil arching, and redistribute the vertical loads. The reinforcement thereby reduces differential settlement and cracking of the soil layer and, in some cases, prevents failure of the soil layer should the foundation collapse locally.

Applications—Typical nonuniform foundations (see Table 1) include: soils with soft, compressible lenses; rock masses containing solution cavities; soils with hard, low compressibility inclusions such as concrete structures and seasonal ice lenses; soils containing buried structures which have the potential to collapse, such as pipes and culverts; and heterogeneous solid waste deposits. Specific examples of geotextile and/or geogrid reinforcement of soil layers built on nonuniform foundations include: clay liners, placed at the bottom of a landfill or surface impoundment, built on a foundation containing sinkholes or soft lenses (Fig. 5*a*); clay covers, constructed on top of solid waste, to close a landfill (Fig. 5*b*); aggregate base course of a road, to prevent catastrophic collapse of the road should solution cavities develop in the rock subgrade after road construction (Fig. 5*c*).

Load Supporting Pads

Construction—One or several layers of reinforcement are typically placed under and/or in compacted fill layers required to support heavy loads in order to reduce foundation shear distortion and/or prevent bearing capacity failure under the applied loads (Fig. 6).

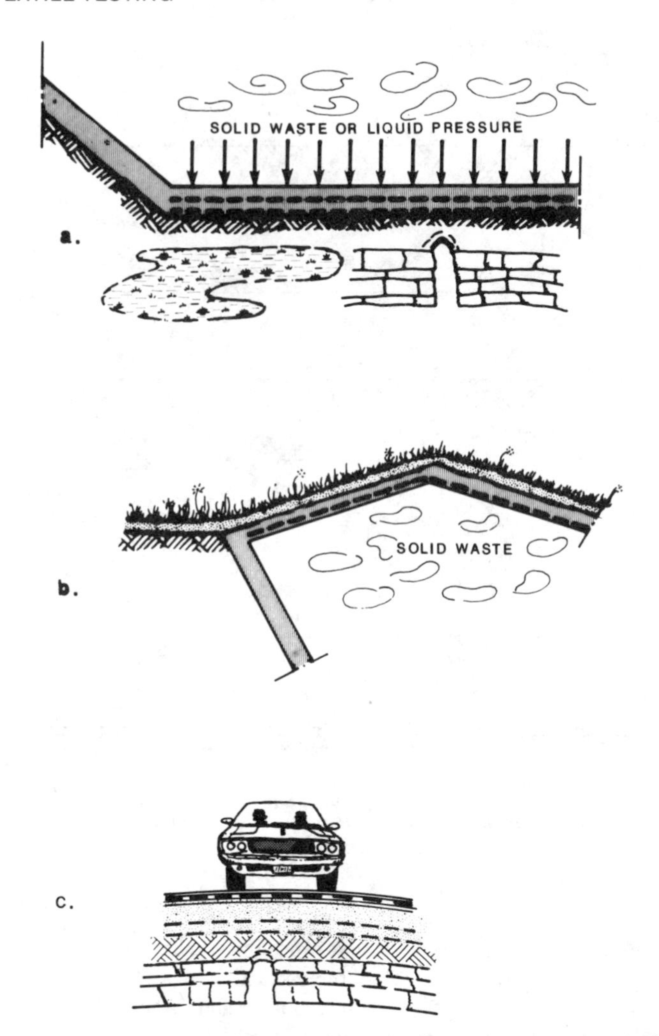

FIG. 5—*Soil layers on nonuniform foundations:* (a) *clay liner at the bottom of a landfill or a liquid impoundment;* (b) *clay cover on top of a landfill; and* (c) *road base.*

Role of the Reinforcement—The reinforcement reduces tensile strains at the bottom of the compacted fill layer and, through shear stress transfer to the foundation soil, mobilizes the shear strength of a larger volume of foundation soil than would be the case for an unreinforced load supporting pad. As a result, the bearing capacity of the compacted fill/foundation soil system is increased, and differential settlements are reduced.

Applications—Heavy loads likely to justify the use of a reinforced load supporting structure (see Table 1) are: large cranes used in fabrication yards, construction staging areas, and transportation intermodal facilities (that is, areas where containers are transferred from trucks to rail cars or vice versa); oil and gas exploration and drilling pads; access roads for exceptionally heavy vehicles. (Note: These heavy vehicles are usually slow, and there is more concern over adequate bearing capacity than over road deterioration caused by repeated traffic and rutting, as is the case for roads and railroad track structures.)

FIG. 6—*Load supporting pads.*

Roads and Railroad Track Structures

Construction—One or more layers of reinforcement are incorporated at various levels in the road structure, either at the base/subgrade interface, in the base layer, or in the asphalt surface course. In railroad track structures, reinforcement is placed at the ballast/subgrade interface or within the ballast layer.

Role of the Reinforcement—In all cases, the reinforcement is intended to reduce lateral and vertical aggregate and subgrade movements caused by traffic loads. Such movements cause cracking in pavements and progressive degradation of the aggregate base layer or railroad ballast. Some types of high tensile modulus reinforcement may also increase the stiffness of the base layer, enabling it to better distribute traffic loads over the subgrade soil and thereby slowing progressive deformation and degradation of the subgrade as a result of repeated loading. Finally, in some unpaved roads where deep rutting is acceptable, the reinforcement increases the bearing capacity of the road structure through a tensioned membrane effect.

Some reinforcement materials placed at the base/subgrade or ballast/subgrade interface also separate aggregate or ballast and subgrade, thereby preventing penetration of aggregate or ballast into the subgrade soil and slowing the intrusion of subgrade soil particles into the aggregate or ballast. As a result, progressive reduction in the "effective" ballast or aggregate layer thickness is prevented, progressive degradation of ballast or aggregate due to fouling is reduced, and, consequently, the long-term behavior of the structure is improved. This "separation" benefit associated with geotextiles and related products was discussed in other papers presented at this symposium.

Applications—Typical applications for reinforcement in these types of structures (Table 1) include:

1. Aggregate bases with a layer of reinforcement at the base/subgrade interface (Fig. 7*a*) used for: unpaved roads (that is, roads without an asphalt or concrete surface layer) such as access roads, haul roads, or "low volume" rural roads; and unpaved areas used as parking lots, storage yards, working platforms, etc.
2. Flexible asphalt pavements (Fig. 7*b*), where a layer of reinforcement may be used to slow

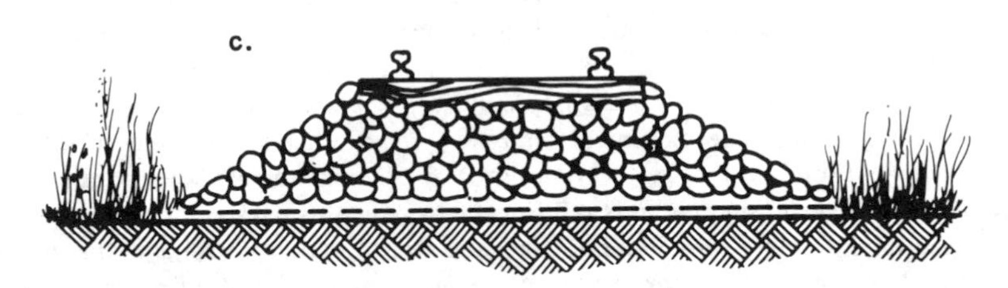

FIG. 7—*Roads and railroad track structures:* (a) *unpaved haul or access road;* (b) *flexible asphalt pavement with aggregate base course; and* (c) *railroad track structure.*

the development of ruts and cracks in the asphalt surface layer and the development of reflective cracks in asphalt overlays. (Note: Geotextiles have been extensively used between existing cracked pavements and pavement overlays to delay crack propagation from existing pavement to overlay; however, as traditionally used, the geotextile may not act as reinforcement but rather as a cushion, isolating the overlay from the existing pavement, and as a water barrier, since the geotextile's permeability is significantly reduced as a result of its impregnation by asphalt during the placement process.)

3. Flexible asphalt pavements (Fig. 7*b*) where one or several layers of reinforcement are placed in the aggregate base course to reduce lateral and vertical aggregate and subgrade displacements, thereby slowing the progressive degradation of the road structure.

4. Railroad track structures (Fig. 7*c*) where a layer of reinforcement (typically a geogrid with large apertures) is used to interlock with and confine ballast rock, thereby reducing lateral ballast spreading.

Fundamentals of Soil Reinforcement

Categories of Soil Reinforcement

Polymeric soil reinforcement materials can be divided into two categories: microreinforcement and macroreinforcement.

Microreinforcement is achieved by mixing into the soil small, usually randomly oriented reinforcing elements such as staple fibers, filaments, yarns, and minigrids. Individual microreinforcing elements influence a volume of soil that is small compared to the total volume of soil contained in the reinforced soil structure. Therefore, a very large number of reinforcing elements is needed. The amount of reinforcement is typically less than 1%, by weight, of the amount of soil. Because the reinforcing elements are small and have large surface areas, they are in contact with many individual soil particles. This is analogous to soil stabilization with admixtures such as cement, lime, or other chemicals where, ideally, every soil particle is in contact with the stabilizing agent. However, there is a difference with respect to the fundamental mechanism of improvement since soil strengthening associated with admixtures is due to chemical reactions which result in a soil cohesion, while microreinforcement is based on mechanical interaction (that is, friction and passive resistance).

Macroreinforcement is achieved by placing into the soil elements that are large compared to the soil particle size. These elements include strips, bars, grids, and fabrics. Individual macroreinforcing elements influence a volume of soil that is significant compared to the volume of the reinforced soil structure. Therefore, a limited number of reinforcing elements is needed. For example, the number of macroreinforcing elements varies from one or two for embankments on weak foundations, up to 20 or more for large reinforced soil walls.

In this paper, only macroreinforcement applications are discussed. Soil reinforcement using fibers, yarns [*1*], or minigrids [*2*] is still in the developmental stage, and these materials are not widely used at present.

Soil Mass-Reinforcement Geometry

Two aspects of soil mass-reinforcement geometry should be considered.

First, the reinforcement can be either localized or distributed throughout the earth structure (that is, the reinforcing elements can be either small in number and placed in specified locations, or large in number and placed throughout the entire considered soil mass). Examples of structures with localized reinforcement are embankments resting on weak foundations with a single layer of reinforcement at their base (see Fig. 4*a*). Examples of structures with distributed reinforcement include reinforced slopes and retaining walls (see Figs. 1 and 2). The distinction between localized and distributed reinforcement has an impact on design.

The second aspect of soil mass-reinforcement geometry concerns only the case where reinforcement is distributed throughout the earth structure. If the distribution of reinforcement has a certain degree of regularity (for example, parallel closely spaced macroreinforcing elements with constant spacing or spacing varying progressively with depth), it may be possible to model the reinforced soil (which is a composite material) as an equivalent continuum [*3*]. This "equivalent continuum" is anisotropic, having a maximum tensile strength in the direction of reinforcement. However, if the reinforcement spacing is large, or if its distribution is too variable to allow evaluation of average properties, it will be impossible to define an equivalent continuum.

Approach to Design

Structures with localized reinforcement, such as embankments on weak foundations, are almost always designed by adding the contribution of the reinforcement to an otherwise classical design approach. This method is called the discrete approach and will be described in following paragraphs.

Structures with distributed reinforcement include those with microreinforcement as well as those with distributed macroreinforcement. The design of these structures considers the external stability of the entire structure acting as a monolith as well as the internal stability of the reinforced soil mass. External stability is usually checked using classical limit equilibrium methods. For internal stability, two approaches to design are possible: the global approach and the discrete approach.

The Global Approach—The global approach involves analysis of the behavior of an equivalent continuum having definable mechanical properties. For example, the strength of a reinforced soil can be evaluated in terms of a Mohr-Coulomb failure criterion with a friction angle similar to the nonreinforced soil but with a higher cohesion. This cohesion may be more or less isotropic in the case of microreinforcement and highly anisotropic for the case of parallel macroreinforcement.

The global approach to design is possible if the actual composite material (that is, the reinforced soil) can be described as an equivalent continuum, if the properties of this equivalent continuum can be evaluated, and if the boundary conditions are known and can be modeled. This approach is well-suited for soils reinforced with microreinforcement since measurement of the properties of these soils can be performed on samples of relatively small size (for example, sample sizes on the order of ten times the distance between the individual microelements). Also, since the mechanical properties of soil with microreinforcement are often more or less isotropic, computations can be made using classical methods based on the mechanics of continua. The global approach is therefore used almost exclusively for this class of problems.

The use of the global approach for soil structures reinforced with macroreinforcement is questionable. It is theoretically possible when an equivalent continuum can be defined. But in order to measure these materials' mechanical properties, the required size of a representative sample must be large (that is, at least five to ten times the spacing between two adjacent reinforcing elements). Since spacing between macroreinforcing elements is usually at least 0.15 m (6 in.), the dimensions of representative samples must be at least 1.5 m (5 ft). Testing such large samples would be very impractical for routine purposes. In addition, the design of testing equipment to apply the appropriate boundary conditions would be very complex. Further complexity is added by the fact that, since macroreinforcement is almost always planar, the representative sample and the reinforced soil mass are highly anisotropic. Consequently, the internal stability of macroreinforced soil masses is usually investigated by a direct analysis of the stress transfer between soil and reinforcement.

The Discrete Approach—The discrete approach consists of analyzing stress transfer between soil and reinforcement. With this approach, classical soil mechanics design methods can be used by incorporating the reinforcement effects into the appropriate equations. This approach can be used for all types of analyses including limit equilibrium analyses (for example, slope stability analyses) and stress-deformation analyses (for example, finite-element analyses).

The discrete approach is almost always used for macroreinforced soil structures because of the difficulty in evaluating the properties of the macroreinforced soil material needed for the global approach. For microreinforced soil structures, the discrete approach is difficult (although theoretically possible using statistical analyses) because of the great number of reinforcing elements and their complex distribution.

Factor of Safety

Traditional factors of safety for unreinforced soil structures are, by and large, empirical. They typically provide a reasonable margin of safety against collapse of the structure under the maximum anticipated loads to which the structure will be subjected. Further, they usually result in limited earth structure deformation so that serviceability criteria are often automatically met. These empirical safety factors incorporate all of the uncertainties about structure and foundation properties, loadings, design assumptions, etc.

There are two commonly used approaches for incorporating the factor of safety into limit equilibrium design equations. The factor of safety can be defined as the ratio of the resistance (forces or moments) of the soil structure and foundation to the applied loading effects. This approach is usually used in design methods that treat the unstable soil zone as a rigid body. Alternatively, the factor of safety can be applied to the soil shear strength to produce factored soil shear strength parameters. This is the approach when the soil is considered to behave as a continuum.

These two approaches give answers that are close to each other for cases in which the soil weight and soil shear strength are the major destabilizing and stabilizing forces, respectively. This is typically the case for unreinforced soil structures.

For reinforced soil structures, the reinforcement forces can be large, and the two cited methods for calculating the factor of safety give different answers. In fact, since the soil and reinforcement often exhibit markedly different stress-strain behavior, no meaningful overall factor of safety can be defined for the reinforced soil structure. An approach that can be used is to apply partial factors of safety to each design variable. Partial load factors can be applied to the soil weight, surcharge loads, seepage forces, and other load effects. Partial resistance factors can be applied to the soil shear strength and reinforcement tensile force.

While the partial factor of safety approach permits consideration of the variability and uncertainty associated with all of the design variables, it has several drawbacks: (1) at present, its use is not widespread with U.S. geotechnical engineers; (2) use of partial factors of safety is slightly more complex than the use of one overall factor of safety; and (3) the partial factor of safety approach has not been correlated with past experience.

Based on the authors' experiences, the following recommendations are provided for incorporating an overall factor of safety into reinforced soil design analyses:

1. Use reinforcement tensile properties for design as described subsequently. Do not factor these tensile strength and resistance values with an additional factor of safety.
2. For evaluation of external stability (that is, when the failure surface does not intersect the reinforcing elements), where the reinforced soil mass is assumed to act as a rigid body, define the overall factor of safety as the ratio of resisting forces or moments to the applied forces or moments.
3. For evaluation of internal stability (that is, when the failure surface does intersect reinforcing elements), apply the factor of safety to the soil shear strength, as follows

$$\tan \phi_f = \frac{\tan \phi}{FS} \tag{1}$$

$$c_f = \frac{c}{FS} \tag{2}$$

where

ϕ = angle of internal friction of soil,
ϕ_f = factored angle of internal friction,
c = soil cohesion,
c_f = factored soil cohesion, and
FS = factor of safety.

The just-cited equations apply to both total and effective stress analyses. These recommendations are in keeping with the most commonly used definitions for the factors of safety of unreinforced soil structures.

Reinforcement Materials

Types of Reinforcement Materials

Polymeric materials used for soil reinforcement include geotextiles and geotextile-related products such as geogrids, strips, mats, webs, meshes, and nets. Today, the most commonly used materials are geotextiles and geogrids made of polyethylene, polyester, and polypropylene.

The three main types of geotextiles are wovens, nonwovens, and knitted. There are two main types of geogrids: those manufactured by drawing a perforated polymer sheet in one or two perpendicular directions; and those manufactured by overlapping perpendicular, or nearly perpendicular, polymer strands and then bonding these strands at their junctions. To date, geogrids manufactured using the first method have been the most widely used for soil reinforcement applications. Descriptions of the wide variety of geotextiles available, how they are manufactured, and information about their properties can be found in Refs *4–7*. Descriptions of geogrids can be found in Refs *6–9*.

Properties of Geotextiles and Geogrids

Table 2 [*10*] lists the many properties of geotextiles and geogrids which may, to a greater or lesser degree, be important for soil reinforcement design. Hydraulic, constructibility, and durability properties and design criteria were discussed in other papers in this symposium. Therefore, this paper will concentrate on the mechanical properties relevant to soil reinforcement design.

All of the mechanical properties for soil reinforcement design can be placed in one of two categories: tensile characteristics and soil-reinforcement interaction characteristics. The following discussion of these two categories is limited to long-term loading conditions. A discussion of repeated or dynamic loading is not included.

Measurement of Tensile Characteristics

Tensile resistance is defined as the tensile force in the reinforcement per unit width at a given tensile strain. Tensile strength is defined as a material's maximum tensile resistance. Occasionally, tensile resistance for design will equal tensile strength, but, in most cases, tensile resistance for design will be significantly less than tensile strength. Since in all the applications described earlier the reinforcement is confined in soil, design tensile resistance should ideally be determined on reinforcement samples confined in soil.

In-Soil Tensile Testing—For many polymeric reinforcing materials, soil confinement significantly influences the behavior of the material, compared to its behavior in the unconfined state. This difference is due to the influences of soil compressive stresses and soil-reinforcement interlocking.

Compressive stresses due to soil overburden or surcharge loads can increase the modulus and strength of some types of reinforcement by causing increased friction between individual fibers, tapes, or yarns in the material. For example, McGown et al. [*11*] found that soil confinement significantly (several hundred percent) increased the initial tangent stiffness [reinforcement tensile modulus multiplied by reinforcement cross-sectional area per unit width (kN/m)] of both a needle-punched nonwoven geotextile and a composite geotextile. It is recommended that in-soil

TABLE 2—*Important properties of geotextiles and geogrids for soil reinforcement design*[a] *(adapted with modification from Bell et al., Ref* 10*).*

Constructibility	Strength and stiffness Temperature stability Ultraviolet light stability Flammability Thickness Absorption Puncture resistance Tear resistance Cutting resistance
Durability	Ultraviolet light stability Temperature stability Resistance to chemical attack Wetting and drying stability Resistance to biological attack Abrasion resistance
Mechanical	Tensile strength Tensile stiffness Boundary friction characteristics Pullout behavior Fatigue resistance Creep resistance Relaxation characteristics Seam strength
Hydraulic	Permittivity Transmissivity

[a] NOTE: All may not be important for every application.

testing be considered for the determination of design parameters for reinforcing materials with complex fabric structures or which are compressible (such as needle-punched nonwoven geotextiles).

Soil-reinforcement interlocking results from penetration of soil particles into the plane of the reinforcement. Three examples of interlocking include penetration of fine soil particles into the surface of a needle-punched nonwoven geotextile, sand grains entering the openings of a woven geotextile, and penetration of fine and coarse soil particles into the apertures of a geogrid. The influence of interlocking should also be evaluated through in-soil testing such as pullout tests or confined tension tests.

Wide Strip Tensile Test—Christopher and Holtz [*12*] discussed some of the difficulties associated with the use of confined in-soil tests to determine reinforcement tensile properties. They suggest that for many reinforcement materials, it is possible to obtain an index value for the design tensile strength and stiffness using a wide strip unconfined constant-rate-of-strain tension test. For many reinforcement applications, this index value, adjusted if necessary to consider the effects of time, temperature, and construction site damage, appears to be sufficient for design purposes. The authors do not recommend the use of traditional grab or narrow strip tension tests for the determination of reinforcement properties for design.

A wide width strip test was recently approved by ASTM Committee D35 on Geotextiles, Geomembranes, and Related Products [ASTM Test Method for Tensile Properties of Geotextiles by the Wide Strip Method (D 4595-86)]. The dimensions of the tested portion of the proposed wide strip specimen are 200 mm (8 in.) wide by 100 mm (4 in.) long. The test is performed at a constant rate of strain of 10%/min. Special consideration may have to be given to clamping high strength (>100 kN/m or 600 lb/in.) geotextiles and geogrids so that no slippage occurs and so that the clamps do not cause stress concentrations and induce premature failure of the specimen. Reduction of specimen width is allowed for materials that have been shown not to be influenced by this variable.

Time Effects—All polymeric materials exhibit time-dependent behavior. Creep is the time-dependent deformation of a specimen under constant tensile load (Fig. 8*a*). Stress-relaxation is the time-dependent reduction in tensile load carried by the reinforcement under constant tensile strain (Fig. 8*b*). Both creep and stress relaxation can occur in polymeric reinforcement during the service life of a typical earth structure. These time effects will significantly influence the mobilized tension in and elongation of the reinforcement. Reinforcement creep has been investigated to only a limited extent. Stress-relaxation effects have been studied even less. Reinforcement stress-relaxation is not considered in most practical design problems. Evaluation of reinforcement creep is described in the following paragraphs.

The ASTM D 4595 standard rate of strain is 10% per minute. At this rate of strain, a reinforcing material with an elongation at break of 30% would require a test duration of 3 min. Many earth structures have a design life of 50 years (3×10^7 min) or more, which is up to seven orders of magnitude larger than the duration of the wide strip test. Depending on the type of polymer, method of manufacture, and other factors, the tensile strength of a polymer reinforcing material loaded to failure over 50 years may be significantly less than half its strength when loaded to failure in 3 min. For design of critical permanent earth structures, reinforcement testing of a duration longer than the proposed ASTM standard is required to account for strength loss due to creep.

Very long-term tests utilizing ASTM D 4595 are usually not possible because of the limita-

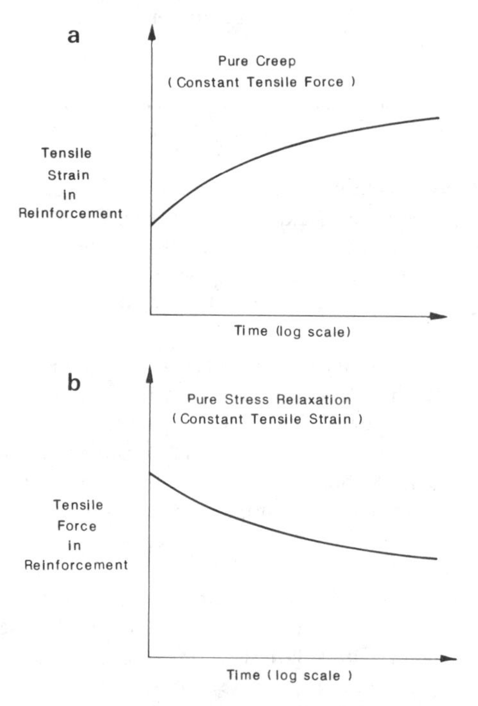

FIG. 8—*Idealized creep and stress relaxation behavior in polymeric reinforcement material:* (a) *creep; and* (b) *stress relaxation.*

tions of most constant rate-of-strain test apparatuses. However, long-term testing can be achieved through a simple modification to ASTM D 4595 wherein the constant rate-of-strain testing apparatus is replaced by a creep testing apparatus consisting of a dead weight and hanger system. In a creep test, the dead weight is quickly and smoothly hung onto one end of the specimen. The other end is attached to a hanger system able to support the applied dead load. The load is maintained for a duration sufficient to extrapolate the test results to the design life of the earth structure. Creep test procedures and their interpretation have been investigated in detail by McGown and his coworkers at the University of Strathclyde [*8*]. Because of the long duration required, material creep testing is not feasible on a project-by-project basis. Therefore, creep test results should be provided by the material manufacturer to the responsible design engineer at the time of design. Results from creep tests can be used to determine time-dependent tensile resistances for design.

In addition to reinforcement creep, the design engineer should be aware of the potential for reinforcement deterioration after long periods of in-ground burial. Deterioration may result from chemical, biological, or mechanical action. Examples of these actions include hydrolysis of polyester in humid environments, environmental stress cracking of polypropylene and polyethylene, polymer oxidation, and microbiological attack. The effects of the just-mentioned processes can be minimized through careful selection of polymer type, polymer characteristics (melt index, density and molecular weight, amount of cross branching, degree of crystallinity, etc.), through the use of products with thick strands (that is, small surface area per unit mass), and through the use of additives such as antioxidants and ultraviolet (UV) stabilizers. Even with these measures, however, the complete elimination of in-ground deterioration is not certain. Additional research is needed to further clarify deterioration mechanisms and to quantify their effects. Manufacturers and researchers should be encouraged to develop material specific "aging factors." These factors would account for any anticipated loss of reinforcement tensile resistance due to deterioration and would be used in the selection of an allowable reinforcement tensile resistance for design of permanent soil structures.

Temperature Effects—To maintain standard laboratory conditions, ASTM requires that geotextile tests be carried out at 21 ± 1°C (70 ± 2°F). This temperature is greater than the in-ground temperature for most applications in much of the United States. Since polymer strength increases as temperature decreases, use of the prescribed test temperature is generally conservative. Examples exist, however, where the reinforcement could be exposed to temperatures well above 21°C (70°F), leading to a substantial reduction in reinforcement tensile strength and modulus. At the other extreme, very cold temperatures result in diminished material ductility and a reduction in elongation to break. Temperature-dependent constant rate-of-strain responses for a polypropylene reinforcing material are shown in Fig. 9. When extreme temperatures will be encountered in the field, the tension test conditions and/or results should be appropriately adjusted.

Construction Site Damage Effects—The effects of construction operations and equipment should be considered in the selection of reinforcement materials and properties for design. Ideally, these properties should be determined by tension testing of reinforcing material specimens that have been subjected to the proposed placement and construction procedures. For example, compaction of blast rock against a geotextile or geogrid may cause damage that reduces these materials' tensile strengths and moduli. In contrast, compaction of beach sand would be expected to have little effect on material properties. In practice, however, it is usually difficult and impractical to test all proposed reinforcement materials for possible damage effects at the time of design (well in advance of construction). In these situations, construction site damage information should be requested from the material manufacturer. Alternatively, an approach suggested by Christopher and Holtz [*12*] can be used. They provide tables for the required degree of geotextile "survivability" (resistance to damage during construction) as a function of subgrade conditions, construction equipment, and type of cover or backfill material. Given the required

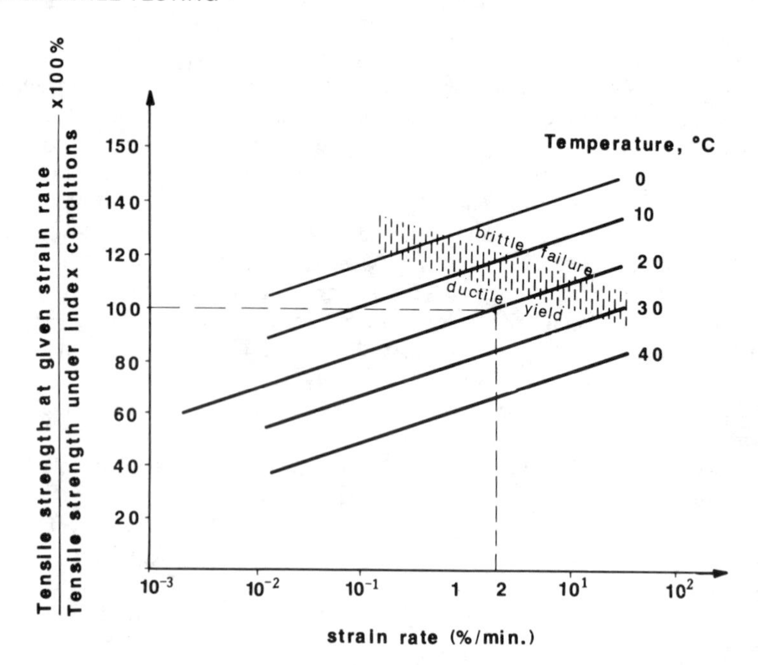

FIG. 9—*Influence of temperature and strain rate: results from constant rate-of-strain tension tests on a polypropylene geogrid. (From McGown et al.* [8].*)*

degree of survivability, the appropriate minimum geotextile properties can be selected from a table developed by Task Force 25 of the AASHTO-AGC-ARTBA Joint Committee on New Materials. This is an interim specification, however, and further research is needed to verify the values given. At present, no such survivability values are available for geogrids. Additional discussion of geotextile survivability can be found in other papers presented at this symposium.

Evaluation of Tensile Characteristics for Design

With due consideration of time, temperature, and construction site damage effects, ASTM D 4595-86 can be used to provide the tensile characteristics required for design. Tensile characteristics relevant to soil reinforcement design include (Fig. 10):

1. *Tensile strength*—the maximum resistance to deformation for a specific material, when subjected to tension by an external force (Point F in Fig. 10).
2. *Tensile resistance*—the force per unit width for a specific material when subjected to a specific strain (Point D is the tensile resistance for Tensile Strain C in Fig. 10).
3. *Initial tangent tensile stiffness*—the ratio of the change in tensile force per unit width to a change in strain along the initial portion of a force per unit width-strain curve (equal to B divided by A in Fig. 10).
4. *Secant tensile stiffness*—the ratio of the change in tensile force per unit width to a change in strain between the origin and any other point on a force per unit width strain curve (equal to D divided by C in Fig. 10).
5. Strain at break (Point E in Fig. 10).

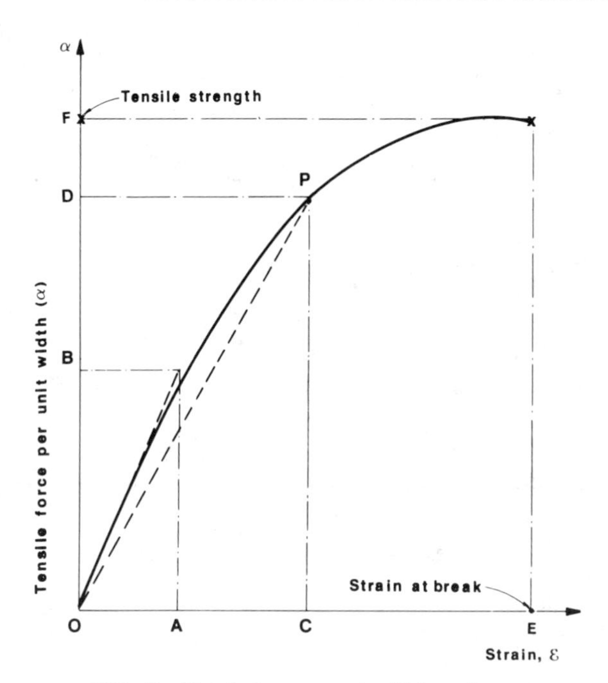

FIG. 10—*Tensile force per unit width-strain curve.*

In critical structures where the reinforcement is expected to permanently resist large tensile loads, the tensile resistance for design should be determined from the results of creep tests rather than from the results of short-term constant rate-of-strain tests. In-isolation creep tests are acceptable for reinforcing products that are not strongly affected by soil confinement. For those products that are strongly affected, in-isolation creep tests will lead to conservative results. Confined, in-soil creep tests, while more difficult to carry out, will result in a more realistic assessment of in-ground performance. Evaluation of the tensile characteristics of polymeric reinforcement from the results of creep tests requires interpretation of the creep test data and extrapolation of the data to the design life of the structure.

Interpretation of creep test data can be carried out using the concept of isochronous (constant-time) curves of tensile resistance versus strain [*8, 9*]. Derivation of isochronous curves from creep test results is illustrated in Fig. 11. Creep tests are performed using a range of tensile resistances. The test results for each tensile resistance (α_i) are first plotted on a strain-log time graph (Fig. 11*a*) and then transposed onto a tensile resistance-strain plot (Fig. 11*b*) to generate an isochronous curve for a constant time (t_i). A set of isochronous curves obtained at a given temperature for a high density polyethylene geogrid is shown in Fig. 12. Once the isochronous curves are established, tensile characteristics can be evaluated for a given design life (given t_i) using the definitions shown in Fig. 10 and described previously. These tensile characteristics should then be adjusted, if judged necessary, to account for nonstandard temperatures, in-ground deterioration, and construction damage.

Methods for extrapolation of polymer creep results are not widely known within the geotechnical engineering community. Information from polymer engineers will be necessary to extrapolate polymer creep behavior to very long load durations. Ward [*13*] and McGown [*8*] discuss methods for extrapolation of creep test data.

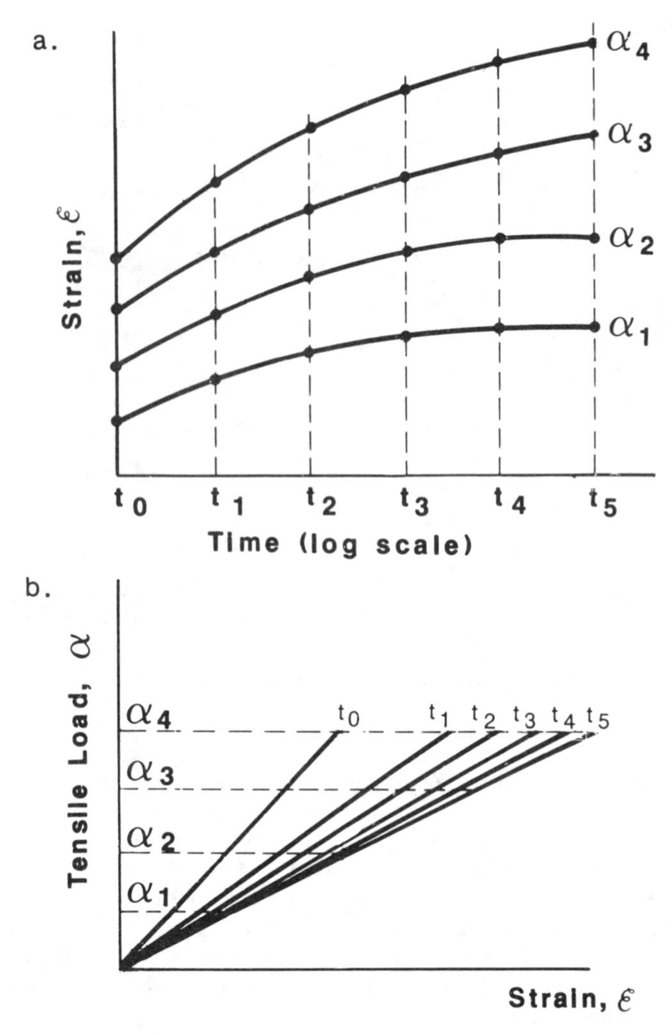

FIG. 11—*Development of isochronous load-strain curves:* (a) *creep test results from tests performed at a constant temperature; and* (b) *creep test results replotted to construct isochronous load-strain curves. (Adapted from McGown et al.* [9].*)*

Soil-Reinforcement Interaction Characteristics

There are two important soil-reinforcement interaction characteristics for design: soil-reinforcement interface shear behavior; and the influence of soil confinement on tensile characteristics. This latter point was discussed in an earlier section.

Soil-reinforcement interface shear behavior can be evaluated using either direct shear tests or pullout tests (Fig. 13). For geotextiles and most geogrids, results from both pullout and direct shear tests are usually presented as a plot of maximum apparent shear stress versus normal stress. The maximum apparent shear stress is calculated from pullout test results by dividing the pullout resistance by twice the embedded surface area of the reinforcement.

Direct shear tests model the condition of soil sliding over the reinforcement. As shown by

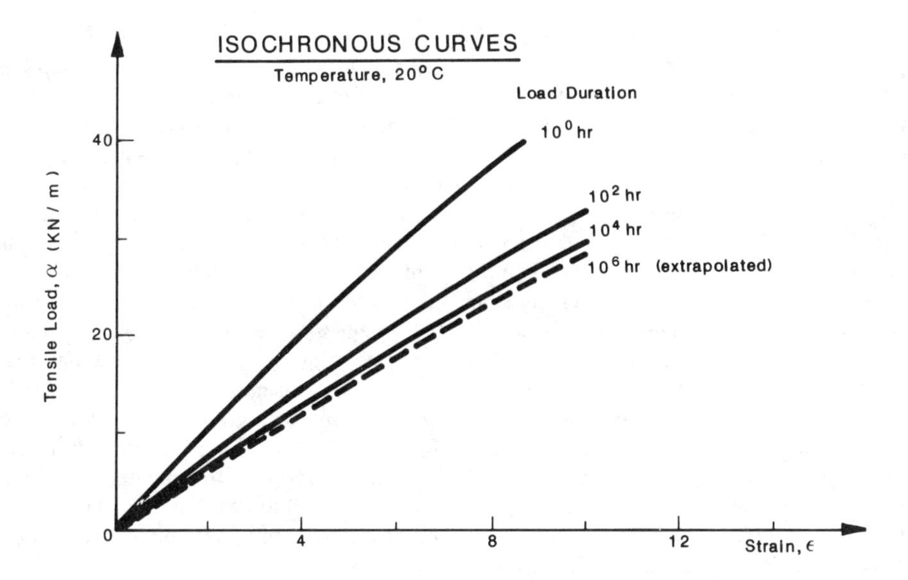

FIG. 12—*Influence of load duration on isochronous load-strain behavior: test results from in-isolation creep tests on a high-density polyethylene geogrid. (Adapted from McGown et al.* [9].*)*

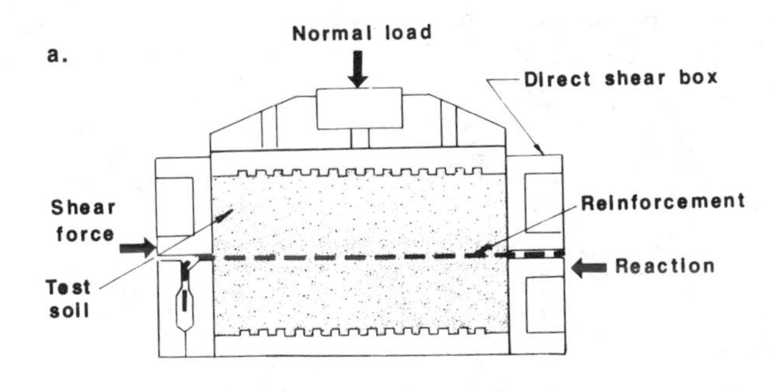

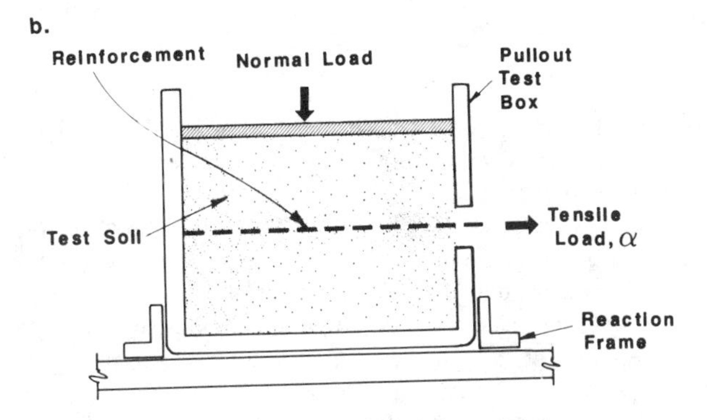

FIG. 13—*Measurement of reinforcement-soil interaction characteristics:* (a) *direct shear test; and* (b) *pullout test.*

Collios et al. [*14*], many external failure mechanisms for reinforced soil structures are of this type. Depending on the relative density, soil type, opening sizes, and surface characteristics of the reinforcement, the angle of interface shearing resistance measured in a direct shear test is less than or at most equal to the angle of internal friction of the soil alone [*12,14,15*]. Pullout tests model conditions associated with reinforcement slip. The pullout test failure mechanism is appropriate for the determination of the required reinforcement embedment length.

Pullout tests have been used for research investigations on extensible polymeric reinforcement [*16–18*]. However, unless appropriate instrumentation is used and special precautions taken, interpretation of the test results is difficult due to complex test apparatus boundary effects and the influence of reinforcement extensibility on the stress distribution along the reinforcement. For these reasons, some researchers [*10,19*] suggest that the direct shear test is more appropriate for design purposes because it is simple to do, requires a less complex apparatus, and yields conservative results. A proposed test method for soil-geotextile friction is currently being developed by ASTM Committee D35. Due to the reasons just given, it is recommended that the direct shear test be used to determine design values for both soil-geotextile sliding friction and geotextile pullout. For geogrids, direct shear tests can be used to evaluate soil sliding, but not pullout. Geogrids have large open areas which allow soil to penetrate the plane of the reinforcement. Due to the large openings, use of direct shear test results for pullout conditions can be misleading and pullout testing will be required.

Soils used for direct shear and pullout tests should be similar to those specified for project construction. Soil compaction conditions in the test should be similar to field conditions. If the reinforcement is at the interface of two different soils, this condition should also be simulated. Ideally, the testing program should account for time and temperature effects, as described previously. Interpretation of the test can be based on peak soil strengths, large-deformation soil strengths, or working conditions. Pullout resistance for geotextiles can be estimated from the results of direct shear tests by

$$T_p = 2(\tan \phi_{sg})\sigma_v L \tag{3}$$

where

T_p = pullout capacity of geotextile per unit width (kN/m),
ϕ_{sg} = angle of soil-geotextile friction,
σ_v = vertical stress (kN/m^2), and
L = length of reinforcement embedded in pullout zone (m).

Reinforced Slopes

Performance Criteria

Reinforcement is used to construct slopes steeper than would be possible without reinforcement. For instance, unreinforced sand is stable on slopes up to about 30 to 40°. Internal reinforcement of a sand mass will permit it to stand at angles up to 90°.

The primary performance criterion for a reinforced slope is adequate stability against sliding of the soils comprising the slope. Sliding can take place either within the reinforced soil mass (internal instability, Fig. 14*a*) or outside the reinforced soil mass (external instability, Fig. 14*b*). The parameter used to define the relative stability of a slope is the factor of safety. For slope stability analyses, the factor of safety is most often applied to the soil shear strength, as discussed previously. Typical factors of safety for slopes range from about 1.2, for small noncritical applications where the consequences of failure are minor, to about 2.0, for high slopes where the consequences of failure may be catastrophic.

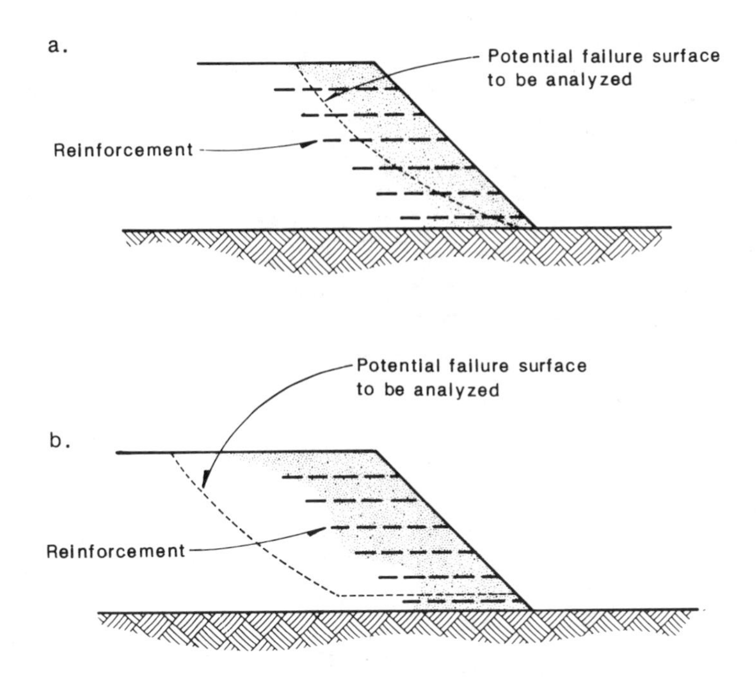

FIG. 14—*Assumed failure modes for reinforced slopes:* (a) *internal failure;* (b) *external failure.*

A second performance criterion, applicable only to slopes supporting roads, buildings, or other structures sensitive to movements, is a limit on slope deformation. For slopes supporting structures, the designer must ensure that the movements required to mobilize the working shear resistance of the soil and the working tensile resistance of the reinforcement are within the permissible limits of the structure being supported. Practically, slope deformations are limited through the use of well-compacted granular fill and suitably large factors of safety.

Design Principles

Methods of analysis which can be used to evaluate the internal stability of a slope include limit equilibrium analyses and stress-deformation analyses. Limit equilibrium analyses are used to evaluate the factor of safety against slope failure. Stress-deformation analyses can be used to evaluate working conditions, although their use in slope design is rare. Therefore, the following discussion is confined to limit equilibrium analyses. It should be pointed out that while limit equilibrium procedures are attractive due to their simplicity and connection to classical design, they involve a number of arbitrary assumptions whose verification through comparison with field performance or detailed numerical studies is extremely limited.

Internal Stability—Internal failure may result from reinforcement rupture (Fig. 15*a*), reinforcement pullout (Fig. 15*b*), or a combination of both. Reinforcement rupture can occur if the tensile force required to maintain equilibrium at any elevation within the slope exceeds the available tensile strength of the reinforcement. Reinforcement pullout can occur if the frictional and/or passive resistance forces developed along the length of reinforcement behind the failure surface are less than the tensile forces required to maintain equilibrium.

Reinforcement forces are incorporated into limit equilibrium slope stability analyses in a number of ways. As shown in Fig. 16, these include: (1) the reinforcement is assumed to act as a

a.

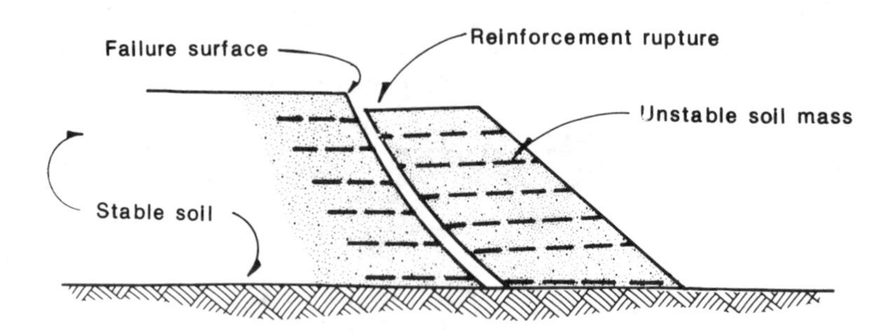

b.

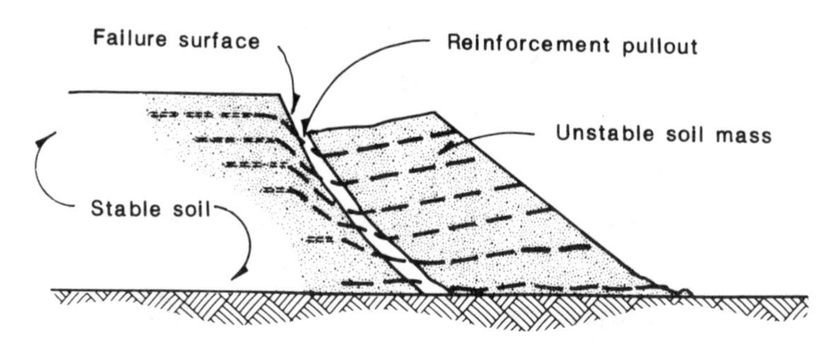

FIG. 15—*Internal failure of reinforcement in slopes:* (a) *reinforcement rupture; and* (b) *reinforcement pullout.*

free body tensile force which does not affect soil strength but which contributes to force and moment equilibrium (Fig. 16*a*); or, (2) the reinforcement is assumed to increase the strength of the slope fill (Fig. 16*b*). With the second assumption, the reinforcement force is usually decomposed into vector components normal and tangent to the slip surface. The component of force parallel to the slip surface is assumed to provide a "pseudo-cohesion," which acts in addition to any soil cohesion. The component of force normal to the slip surface is assumed to increase the normal stress acting on the soil and thereby increase the soil's shearing resistance due to the frictional component of shear strength. With either of the just-cited approaches, the reinforcement orientation at failure can be assumed to be horizontal or it can be assumed to be inclined due to localized large deformations in the vicinity of the failure surface. The maximum possible amount of reinforcement reorientation would result in a reinforcement direction parallel to the slip surface.

A comparison of the two assumptions outlined in Fig. 16 shows that, for circular failure surfaces and horizontal reinforcement orientation, Fig. 16*a* is more conservative (the reinforcement causes smaller stabilizing moments) than Fig. 16*b*. If the reinforcement is assumed to have an orientation at failure parallel to the slip surface, Fig. 16*a* and Fig. 16*b* give identical

a.

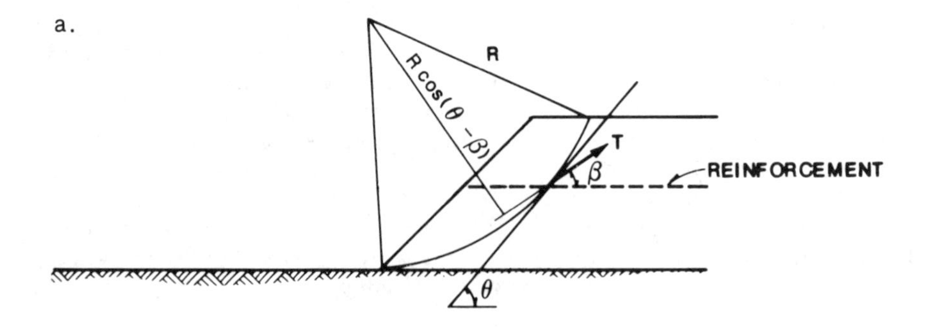

$$M_T = T \cdot [R \cos(\theta - \beta)]$$

$$0 \leq \beta \leq \theta$$

b.

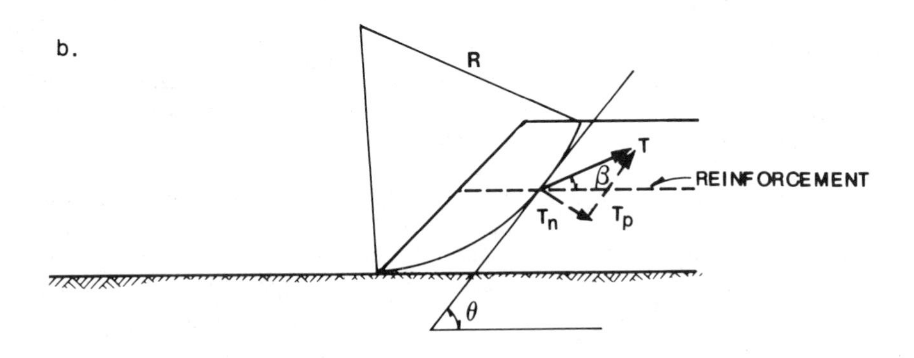

$$M'_T = T_p \; R + T_n \tan \phi' \cdot R$$

$$= RT \, [\cos(\theta - \beta) + \sin(\theta - \beta) \tan \phi']$$

$$= M_T [1 + \tan(\theta - \beta) \tan \phi']$$

FIG. 16—*Stabilizing moment (M_T) due to reinforcement force:* (a) *reinforcement force assumed to act as an independent free-body force which does not effect soil strength; or,* (b) *reinforcement force assumed to increase soil strength.*

results which are less conservative than those for horizontal orientation. Based on the authors' experiences and judgement, the use of Fig. 16*b* along with a horizontal reinforcement orientation is recommended for the design of slopes reinforced with multiple layers of regularly spaced reinforcement. (Note: This recommendation is different than one that will be made for the design of reinforced embankments on weak foundations.)

External Stability—External stability can be evaluated using the same slope stability analyses used to evaluate internal stability. The potential failure surfaces to be investigated should include those passing behind the reinforced soil mass, through the foundation soil, and along the interface between the reinforced soil mass and the foundation soil.

Reinforcement Orientation and Length—Ideally, reinforcement should be placed in the direction of maximum tensile strain. In slopes resting on competent foundations, the principal compressive strains are nearly vertical and the principal tensile strains are nearly horizontal [*20*]. Therefore, placement of the reinforcement in horizontal layers provides a high degree of reinforcement efficiency. Furthermore, the reinforcement is almost always placed horizontally because of the ease of construction.

Reinforcement should be long enough to encompass the entire unreinforced soil mass having potential failure surfaces with a factor of safety smaller than the specified design value. Using this approach, internal and external stability criteria are automatically satisfied. A final check should be made to verify that the reinforcement length is adequate to prevent reinforcement pullout.

Simplified Design of Reinforced Slopes

Simplified design charts have been developed for reinforced slopes [*21–24*] assuming a bilinear sliding wedge failure mechanism and a cohesionless slope fill. These charts can be used for the preliminary design of reinforced slopes. Schmertmann et al. [*24*] developed charts for determining the amount, length, and distribution of geogrid reinforcement required to maintain equilibrium in slopes constructed with cohesionless free-draining fills resting on competent, level foundations. The charts, which are extensions of earlier charts by Jewell et al. [*22*], were developed by evaluating the reinforcing force required to maintain horizontal equilibrium in the slope, as shown in Fig. 17. The angle of the resultant force between slices was assumed to be equal to the mobilized soil friction angle (which is assumed equal to the factored soil friction angle given by Eq 1: $\delta = \phi_f$). The reinforcement was assumed to act as shown in Fig. 16*b* and have a horizontal orientation at failure. The coefficient of friction for soil sliding over reinforcement was assumed to be 90% of the coefficient of friction for soil sliding over soil; this value is appropriate for granular soil sliding over geogrids, but it may not be appropriate for some geotextiles or for cohesive soils. The use of Schmertmann et al.'s charts are described in following paragraphs.

Amount of Reinforcement—The total horizontal tensile reinforcement force (T) per unit width of slope required to maintain equilibrium in a slope is calculated using Fig. 18. T is the sum of the required tensile forces at all reinforcement levels. The chart provides a force coefficient (K), derived from the results of the two-part wedge analyses (Fig. 17), which is used to calculate T (kN/m) as follows

$$T = 0.5K\gamma H^2 \tag{4}$$

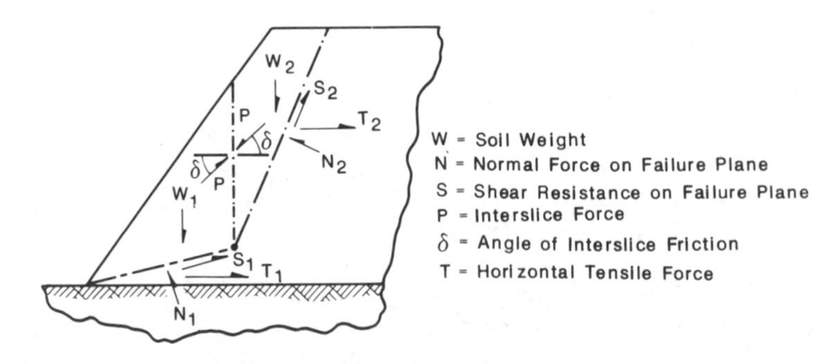

FIG. 17—*Two-part wedge failure mechanism. (From Schmertmann et al.* [24].*)*

where

K = dimensionless coefficient given by Fig. 18,
γ = unit weight of soil (kN/m^3), and
H = height of reinforced slope (m).

The input parameters to use with Fig. 18 are: β = slope angle; and ϕ_f = factored soil angle of internal friction, obtained using Eq 1 so as to incorporate the design factor of safety.

The required minimum number of layers of reinforcement, N, is determined by

$$N = T/\alpha_a \tag{5}$$

where

α_a = allowable tensile resistance provided by the reinforcement. Selection of the allowable reinforcement tensile resistance will be discussed subsequently.

The effect of a surcharge load, q (kN/m^2), uniformly distributed on top of the slope, can be estimated by assuming that the surcharge loading is equivalent to an additional thickness of soil

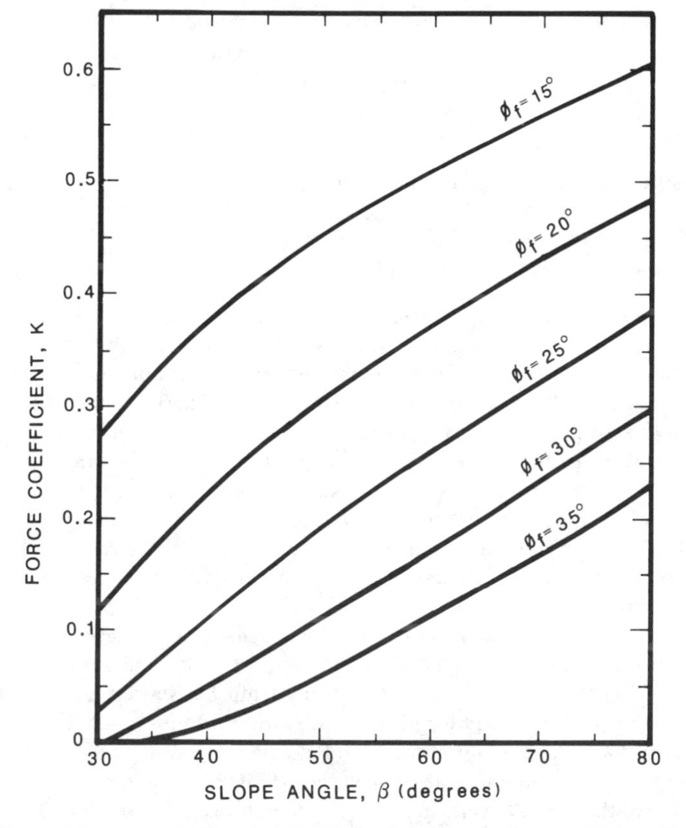

FIG. 18—*Slope reinforcement design: chart for determination of the force coefficient,* K, *for the design of reinforced slopes constructed with granular-free draining fill. (Note:* β = *angle slope makes with horizontal;* ϕ_f = *factored soil friction angle; from Schmertmann et al.* [24].*)*

placed on top of the slope. This equivalent thickness of soil results in an equivalent slope height, H' (m), given by

$$H' = H + q/\gamma \tag{6}$$

where

H = actual height of reinforced slope (m), and
γ = unit weight of soil (kN/m^3).

The equivalent slope height is used to calculate the total required horizontal tensile reinforcement force (Eq 4) as well as the reinforcement length. The use of the equivalent slope height concept is only applicable if the equivalent thickness caused by the surcharge (q/γ) is small compared to the actual slope height (H).

Distribution of Reinforcement—The next design step consists of distributing the N layers of reinforcement in a way that ensures stability of the reinforced slope at every level. (If all the required reinforcement was put at one level, Eqs 4 and 5 would be satisfied but the portions of slope above and below would be unstable.) The amount of reinforcement required at each level is proportional to the rate of change of total required tensile force with depth, $dT/dz = K\gamma z$, as shown by Jewell et al. [22]. The required amount of reinforcement therefore increases linearly with depth. Consequently, the maximum spacing S_v (m) between reinforcement layers will decrease inversely with depth, z (m), below the crest of the equivalent slope, according to:

$$S_v = \alpha_a/(K\gamma z) = H'^2/(2Nz) \tag{7}$$

where

K = dimensionless tensile force coefficient given by Fig. 18,
γ = unit weight of soil (kN/m^3),
H' = equivalent height of slope (m), and
N = minimum number of layers of reinforcement.

The selected spacing should, of course, be compatible with practical soil compaction lift thicknesses. Usually, to achieve a reinforcement layout compatible with compaction lift thicknesses, the final number of layers of reinforcement will exceed the minimum, N, by one or two.

If the number N is small, reinforcement layers may be far apart and there is a risk of slope failure between the reinforcement layers. From experience, a maximum spacing value of 1.0 m (3 ft) is recommended. If the number of layers, N, given by Eq 5 is such that spacing between reinforcement layers is larger than this recommended maximum, then either 1.0 m (3 ft) should be retained as the spacing or another type of reinforcement, with a smaller value of α_a, should be considered. The latter alternative is usually the more economical. A third alternative is to add a few intermediate layers of a weaker reinforcement material to improve stability between layers of main reinforcement.

Length of Reinforcement—The length of reinforcement is determined using Fig. 19. The chart shown in this figure was developed using the two-part wedge analysis indicated in Fig. 17 and is based on two criteria: (1) the reinforcement length must be sufficient to prevent reinforcement pullout; (2) the reinforcement length must be sufficient to prevent outward sliding along the interface of the soil and the bottommost reinforcement layer.

Figure 19 provides a minimum reinforcement length at the bottom of the slope, L_B(m), as well as at the top of the slope, L_T(m). In all cases, the minimum length of reinforcement at the bottom of the slope is larger than or equal to the minimum length at the top. A linear variation of reinforcement length between the top and bottom values is appropriate for reinforcement at intermediate elevations.

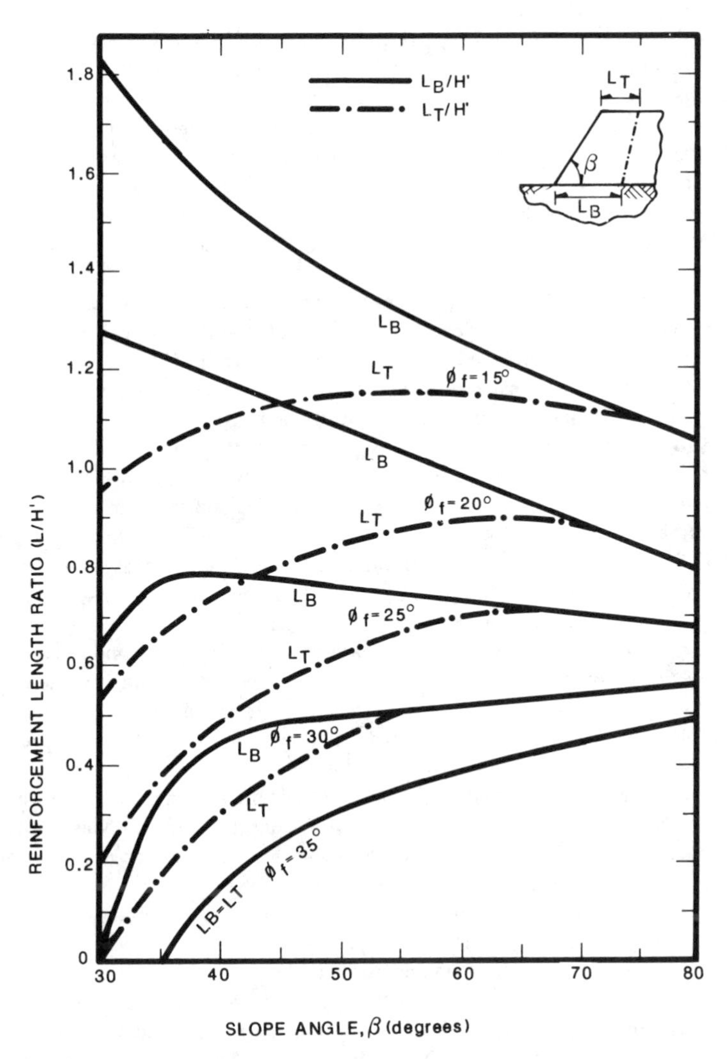

FIG. 19—*Slope reinforcement design: chart for the determination of reinforcement length,* L, *for slopes constructed with granular-free draining fill. (This chart is for a soil-reinforcement interface shear strength equal to 90% of the soil shear strength.) (Note:* ϕ_f = *factored soil friction angle;* H′ *is given by Eq 6; from Schmertmann et al.* [24].*)*

Reinforcement Properties and Relevant Test Methods

Limit Equilibrium Design—Ideally, the allowable reinforcement tensile resistance for design of reinforced slopes should be based on long-duration confined creep tests. Unfortunately, data from this type of test are virtually nonexistent. In fact, unconfined creep test results are available for only a few commercially available geotextile and geogrid products. For critical applications, allowable tensile resistances should be based on creep tests and specifying engineers should require certifiable creep test data from the material manufacturer as a prequalification for the specification of a material for a critical structure. For less critical applications, a simpler approach is justified: allowable tensile resistances can be based on the ASTM wide strip test

[ASTM Test Method for Tensile Properties of Geotextiles by the Wide Width Strip Method (D 4595-86)] conducted at the standard rate of strain.

Selection of an allowable tensile resistance from the wide strip test should consider the limiting allowable deformations of the reinforced soil structure, as well as the effects of reinforcement-soil interaction, time (creep and in-ground deterioration), temperature, and construction damage, as outlined earlier. The approach most often taken is to select an allowable tensile resistance significantly less than the reinforcement tensile strength measured in the constant rate-of-strain test and to minimize the effects of construction damage through specification of a suitably robust material.

Limit equilibrium analyses for slopes can be carried out using peak soil shear strengths or large-strain, constant volume ("critical state") soil shear strengths. In well-compacted fills, the peak soil shear strength is developed at just a few percent strain. If this peak is used in design, the allowable reinforcement tensile resistance should be obtained from the wide strip test results at a compatible strain value (a range of 2 to 5% strain may be appropriate for design). When the large-strain, constant volume soil shear strength is used, the allowable reinforcement tensile resistance may be established at a larger value of strain (a range of 5 to 10% strain may be appropriate for design). In all cases, it is suggested that the allowable tensile resistance be limited to not more than 20 to 40% of the peak tensile strength measured in the ASTM wide strip test (D 4595-86) to account for the uncertainties associated with the effects of time, temperature, and construction damage. The actual percentage of the wide strip tensile strength to use in design should be established on a product specific basis.

For seismic design and dynamic loading situations, wide strip tests conducted at higher than standard rates of strain may be considered. This will often (depending on the polymer from which the reinforcement is manufactured and the physical structure of the reinforcement) result in an allowable tensile resistance for seismic and dynamic loading conditions appreciably higher than the corresponding sustained loading resistance. As shown by Bonaparte et al. [*25*], the combination of (1) the increase in allowable tensile resistance of polymeric reinforcing elements during high strain-rate loading and (2) the low allowable factor of safety associated with rare earthquake events results in a design for seismic conditions that is often automatically satisfied by the long-term sustained loading design.

Deformations Under Working Conditions—The just-cited recommendations on allowable reinforcement tensile resistances are for limit equilibrium analyses. Since the factor of safety of the reinforced slope will be greater than one, the actual stresses and strains in the reinforcement will be lower than those corresponding to the limit state condition. The tensile strain under actual working conditions can be understood by considering the load-strain characteristics of the soil and the reinforcement, as outlined in Fig. 20. It can be seen that the tensile strain under working conditions (ϵ_1) in the reinforcement shown in Fig. 20 is less than the strain corresponding to the tensile resistance used in design (ϵ_2 if the design is based on the peak soil strength and ϵ_3 if the design is based on the large-strain, constant volume soil strength). It should also be remembered that the distribution of strain along the length of a reinforcing element varies from zero at its ends to a maximum at the point of maximum tensile force. Under working conditions, the strain in the reinforcement may be negligible along a significant percentage of the reinforcement's length, and the average working strain in the reinforcement will be significantly less than the maximum working strain. Based on the low average strains under working conditions, the overall structural deformations in a reinforced slope with an adequate factor of safety will be small and will be largely controlled by the stress-strain characteristics of the compacted fill.

To reduce the potential for catastrophic failure due to reinforcement rupture, it is recommended that the tensile strain in the reinforcement at break be at least two to three times the strain required to mobilize the peak strength of the soil. In practical terms, this means that the minimum strain at break should be at least 10 to 15%.

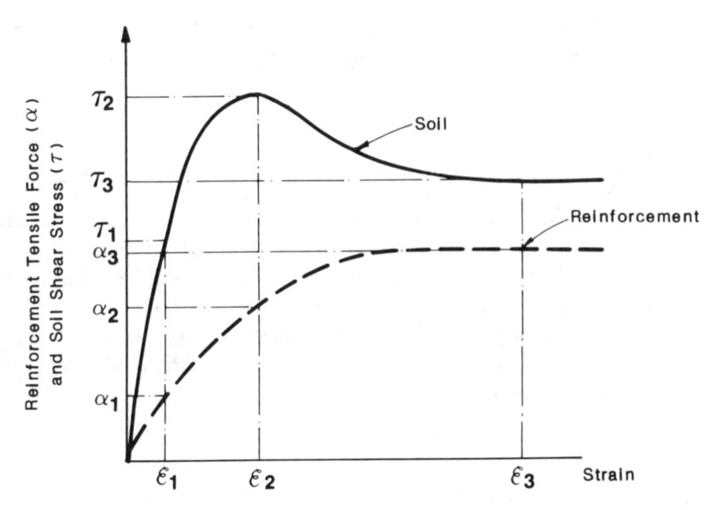

FIG. 20—*Working strains in reinforced slopes: Subscript 1 refers to loads and strains under working conditions; Subscript 2 refers to loads and strains when the soil has reached its peak strength; Subscript 3 refers to loads and strains under large-strain constant volume conditions.*

Soil-Reinforcement Interaction—As outlined previously, both soil-geotextile interface friction and geotextile pullout can be evaluated using results from direct shear tests. For geogrids, direct sliding of the soil over the grid can be estimated using direct shear tests. Pullout resistance should be evaluated using the pullout test. In the absence of specific test data, see Christopher and Holtz [*12*] for a summary of available direct shear and pullout test data. Alternatively, for direct sliding, it can usually be assumed that for compacted cohesionless soils the soil-reinforcement friction is equal to two-thirds of the soil's internal friction.

Reinforced Soil Walls

Performance Criteria

Most reinforced soil walls are used as retaining walls, often supporting a road, railroad track, or bridge structure. Consequently, the two relevant performance criteria are stability and acceptable deformations.

The primary performance criterion for a reinforced soil wall is adequate stability. Failure can occur either within the reinforced soil mass (internal stability) or outside the reinforced soil mass (external stability). The parameter used to define the relative stability of a wall structure is the factor of safety. The factors of safety for internal and external stability are often different. For internal stability, the factor of safety is normally applied to the soil shear strength. For external stability, however, the factor of safety is applied to the stabilizing forces (or moments).

The secondary performance criterion is a limit on wall deformation, which implies a limit on the movements required to mobilize the working shear resistance of the soil and the working tensile resistance of the reinforcement. The limit on wall deformation is selected on the basis of architectural considerations regarding the wall facing, on the visual impact of the wall, and on servicability criteria for supported structures such as pavements and bridges. Since the stress-strain characteristics of the wall fill have a considerable influence on construction-related and postconstruction wall movements, good quality, well-compacted granular fill is often specified for reinforced soil walls. However, recently, a number of relatively low-height walls have successfully been constructed with silty and/or clayey backfill materials.

Design Principles

Ideally, a reinforced soil wall should be designed using a two-step procedure:

1. A limit equilibrium analysis to check on the stability of the wall and to determine the factor of safety against failure. Limit equilibrium analyses for walls use methods identical to those for reinforced slopes. Both external and internal stability should be considered.
2. A stress-deformation analysis under working conditions to check that stresses and strains in the reinforced soil mass are compatible with the properties of the soil and reinforcement, and to evaluate vertical soil settlements and lateral wall movements. (Stress-deformation analyses are more important for walls than for slopes, since walls often have more stringent servicability criteria.)

Simplified Design of Reinforced Soil Walls

At the present time, reinforced soil walls are designed using methods that are not as complete as the ideal two-step procedure just described. Most simplified design procedures consider internal and external stability independently. External stability calculations are carried out considering the reinforced soil mass to be a rigid block. The two most common approaches to the evaluation of internal stability are the semiempirical coherent gravity and the tieback wedge design procedures [*26*]. Both procedures involve solution of relatively simple closed form equations, and both were originally developed for the patented Reinforced Earth wall system (reinforced soil walls with metallic strip reinforcement and articulated segmental concrete panels).

The coherent gravity procedure, first presented by Schlosser [*27*], is a working stress method based on the observed performances of both field and laboratory scale structures. The assumed kinematic mechanism for wall movement is rotation about a hinged crest. The locus of maximum reinforcement tensile forces is assumed to be a two-part surface (Fig. 21*a*). The assumed earth pressure distribution along the two-part surface ranges from an at-rest condition at the top of the wall to an active condition in the lower portions of the structure. The coherent gravity procedure is currently the most commonly used method of design for reinforced soil walls with metallic reinforcement.

The tie-back wedge procedure is a limit equilibrium method of design. It assumes mobilization of the full shear strength of the reinforced fill and generation of active earth pressures along the potential failure surface. The classical Rankine failure surface is assumed to be the locus of maximum reinforcement tensile forces (Fig. 21*b*). The assumed kinematic mechanism for walls movement is rotation about a hinged toe. Lateral earth pressures larger than the active values have also been used with the tie-back wedge failure mechanism. Steward et al. [*28*] used the at-rest earth pressure for geotextile reinforced soil retaining walls. Broms [*29*] suggested the use of an earth pressure coefficient slightly larger than the at-rest value.

Berg et al. [*30*] have discussed the application of both the coherent gravity and tie-back wedge procedures to soil walls reinforced with relatively extensible polymeric materials. They concluded that the horizontal extension required to generate active lateral earth pressures in the reinforced fill is significantly less than the working strain in polymeric reinforcement. Active lateral earth pressures should therefore develop in polymer reinforced walls as long as the wall facing elements are not constrained. They also suggested that the kinematic mechanism associated with the tie-back wedge procedure better reflected the behavior of walls with extensible reinforcement. On this basis they recommend the use of the tie-back wedge procedure and active lateral earth pressures for the design of soil walls reinforced with polymeric materials.

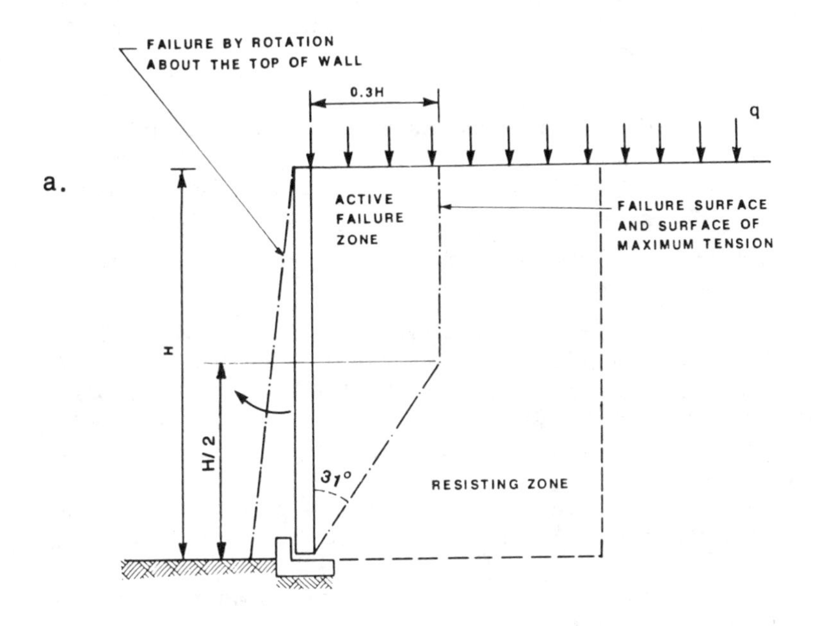

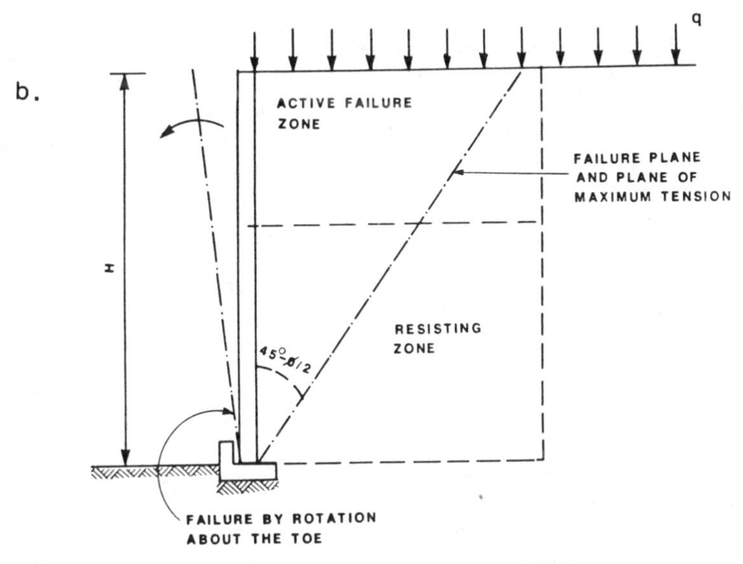

FIG. 21—*Design methods for reinforced soil walls:* (a) *coherent gravity procedure; and* (b) *tieback wedge procedure.*

Internal Stability—Evaluation of internal stability of a reinforced soil wall using the tie-back wedge method involves the calculation of the lateral earth pressures that must be resisted by the reinforcement tensile forces. The lateral earth pressures are assumed to act on the Rankine failure surface which is also assumed to be the locus of the maximum reinforcement force. The type, length, and number of layers of reinforcement are selected to prevent either of two different failure mechanisms:

1. Rupture of the reinforcement along the assumed locus of maximum reinforcement tensile force.
2. Pullout of the reinforcement from the stable resisting zone behind the assumed failure surface.

Evaluation of these two internal failure mechanisms using the tie-back wedge method of analysis is carried out using the following steps:

1. The maximum vertical stress induced at every level in the reinforced soil by gravity, surcharges, and the active thrust from the retained fill (fill behind the reinforced zone) is calculated. Classical soil mechanics methods for stress distribution are used in this step.
2. The maximum horizontal stress versus depth that must be resisted by the reinforcement is obtained by multiplying the maximum vertical stress by the coefficient of active lateral earth pressure, K_a.
3. The required tensile resistance at any reinforcement level is equal to the maximum horizontal stress multiplied by the vertical spacing between reinforcement layers at the considered level.
4. A reinforcing material is selected and, based on its characteristics, a practical reinforcement spacing and layout is determined.
5. The reinforcement length is calculated based on the required length of reinforcement behind the assumed failure surface to prevent pullout.

Internal stability calculations using the assumed Rankine failure surface and the five aforementioned steps can be carried out as follows.

The maximum vertical stresses induced by gravity, uniform normal surcharges, and the active thrust from the retained fill are calculated assuming an internal vertical stress distribution at any Depth z (m), given by Fig. 22:

$$\sigma_{z1} = \frac{\gamma z + q}{1 - \left[\dfrac{K_{ar}(\gamma_r z + 3q)}{3(\gamma_r z + q)}\right](z/L)^2} \tag{8}$$

where

σ_{z1} = maximum vertical stress at Depth z below the top of the reinforced soil wall (kN/m^2),
γ = unit weight of reinforced soil (kN/m^3),
q = uniformly distributed normal surcharge at the top of the reinforced soil wall and retained fill (kN/m^2),
K_{ar} = coefficient of active earth pressure of retained fill,
γ_r = unit weight of retained fill (kN/m^3), and
L = reinforcement length (m).

The reinforcement length, L, used in Eq 8, is obtained from the external stability analysis which is typically carried out prior to the evaluation of internal stability.

Equation 8 has been established with the assumption that active thrust from the retained fill

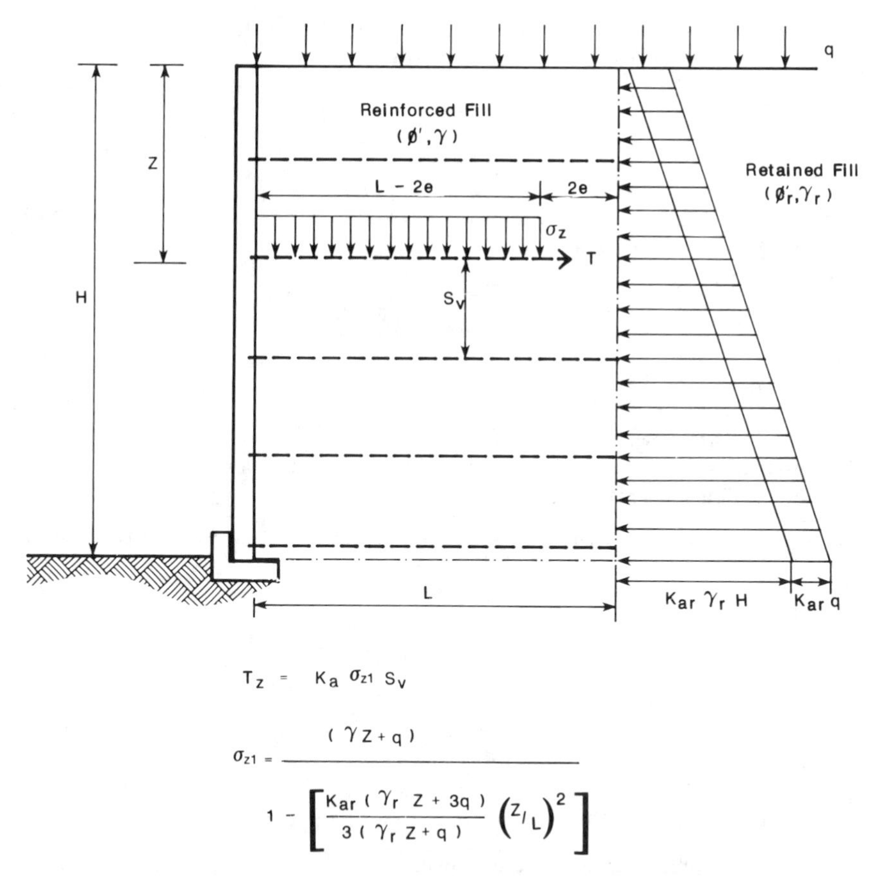

FIG. 22—*Vertical stresses behind reinforced soil wall: stresses due to gravity, uniform surcharge, and active thrust of the retained fill. (The vertical stresses induced by the thrust of the retained fill are calculated here using Meyerhof's recommendations for eccentrically loaded footings. Other assumptions regarding the vertical stress distribution are possible.)*

is horizontal. This is common practice and conservative. Further, use of Eq 8 assumes that the vertical stresses induced by this active thrust are applied in a manner analogous to Meyerhof's stress distribution for eccentrically loaded footings (see Fig. 22). Other assumptions are possible for the distribution of vertical stresses caused by the active thrust of the retained fill [*26,31*].

Maximum vertical stresses σ_{z2} resulting from localized normal loads can be calculated either by using published elastic solutions [*32,33*] or by assuming a 2:1 trapezoidal stress dispersal between the top of the reinforced soil wall and the considered reinforcement elevation [*26*].

The maximum horizontal stress at Depth z due to gravity, normal surcharges, and the thrust of the retained fill is calculated by

$$\sigma_{h1} = K_a(\sigma_{z1} + \sigma_{z2}) \tag{9}$$

where

σ_{h1} = horizontal stress to be resisted by reinforcement at Depth z resulting from gravity, normal surcharges, and retained fill thrust (kN/m), and

K_a = coefficient of active pressure of reinforced soil.

The horizontal stress at Depth z resulting from a horizontal strip surcharge at the top of the reinforced soil wall is calculated directly as follows

$$\sigma_{h2} = (2T/h)[1 - (z/h)] \tag{10}$$

where

σ_{h2} = horizontal stress to be resisted by reinforcement at Depth z resulting from a horizontal strip load T (kN/m), and
h = the lesser of H or h_o given by

$$h_o = [(b + 2d)/2]/\tan(45° - \phi/2) \tag{11}$$

where

b = width of strip (m),
d = distance between wall facing and centerline of strip (m), and
H = total wall height (m).

The required tensile force per unit width in the reinforcement at Depth z is finally given by

$$\alpha = (\sigma_{h1} + \sigma_{h2})S_v \tag{12}$$

where

S_v = reinforcement vertical spacing, m, and
α = force per unit width in reinforcement at Depth z resulting from gravity, normal and horizontal surcharges, and thrust from the retained fill (kN/m). In practical design, the allowable reinforcement tensile resistance is known (a type of reinforcement has been selected), and Eq 12 is used to solve for reinforcement spacing versus depth.

Finally, the required length of reinforcement beyond the Rankine surface to prevent reinforcement pullout is determined by equating T_p, the reinforcement pullout capacity (determined from pullout tests or direct shear tests and Eq 3), to the required reinforcement force in Eq 12. The total length of a layer reinforcement is equal to the length of reinforcement required to prevent pullout (Eq 3) plus the length of reinforcement between the wall face and the potential failure surface. This total length, determined for each reinforcement layer, is then compared to the length obtained from the external stability analysis, and the longer of the two is selected.

External Stability—External stability is evaluated assuming the reinforced soil mass acts as a rigid body which resists external loads, including the earth pressure from the soil being retained behind the reinforced soil mass and loads applied to the top of the soil mass, without failure by one of the following mechanisms:

1. Sliding along the base of the reinforced soil mass or along any plane above or below the base.
2. Overturning about the toe of the reinforced soil mass.
3. Bearing capacity failure of the foundation soil.
4. General slope failure.

These four mechanisms are illustrated in Fig. 23. Classical soil mechanics methods are used for verifying external stability and need not be discussed here. Calculations related to the four just-cited mechanisms show that the external stability is increased if the width of the assumed rigid body is increased, (that is), if the length of reinforcement is increased.

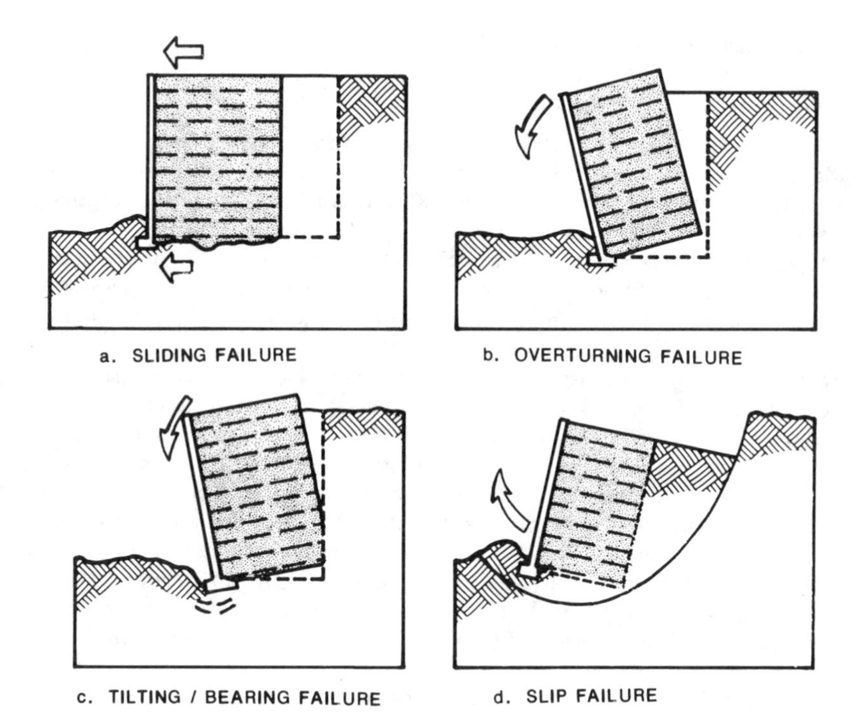

FIG. 23—*External stability of a reinforced soil wall:* (a) *sliding along base;* (b) *overturning;* (c) *bearing capacity failure; and* (d) *general slope failure.*

Reinforcement Properties and Relevant Test Methods

The selection of reinforcement properties and relevant test methods for soil retaining wall design is similar to the selection of properties and test methods for slope reinforcement applications. Several properties deserve special consideration in retaining wall applications, however. These are:

1. *Initial tangent stiffness:* serviceability criteria for walls, particularly concrete panel faced walls, are typically more stringent than those for slopes. Therefore, less strain under working conditions can be tolerated.
2. *Resistance to environmental exposure:* polymer reinforcement connected to thin concrete facing panels or exposed to summer sunlight may reach temperatures in the range of 40°C (100°F). The effects of these temperature levels on reinforcement strength, stiffness, and creep should be considered. If the reinforcing element is used as the facing for the wall, its stability when exposed to UV light must be considered. For temporary walls, UV stability may not be a problem. However, the effects of fire or vandals on the facing may need to be considered. For permanent walls it will often be advisable to cover the exposed face of the wall. Timber, shotcrete, and vegetation have been used to cover geosynthetic wall facings.
3. *Resistance to construction site damage:* fill compaction requirements behind retaining walls are often more stringent than the requirements for slopes. In addition, for permanent highway structures, hard angular quarry rock is commonly used as backfill. Therefore, construction site damage potential should have special consideration on these large permanent retaining wall projects.

Reinforced Embankments on Weak Foundations

Performance Criteria

Geotextile or geogrid reinforcement is used at the interface between embankments and weak foundations to enable the construction of embankments which would otherwise cause failure of the foundation soil. Failure of embankments constructed on weak foundations without reinforcement can be prevented by limiting the height of the embankment and/or the steepness of its side slopes. Reinforcement permits construction of embankments higher or with steeper side slopes than would be possible without reinforcement. Reinforcement is sometimes used in conjunction with other soil improvement measures such as staged construction, side berms, stone columns, and vertical drains.

The two performance criteria usually considered for embankments are adequate stability and acceptable total and differential settlement. Reinforcement placed at the embankment/foundation interface enhances stability and reduces embankment spreading and initial (undrained) embankment settlements. However, for embankments over uniform foundation layers (see Fig. 4*a*), reinforcement usually has little influence on the total settlement of the embankment because it does not significantly modify the vertical stresses exerted by the embankment on the foundation soil. For nonuniform foundation conditions (see Fig. 4*b*), reinforcement at the embankment/foundation interface may significantly reduce total and differential settlements.

The parameter used to indicate the relative stability of an embankment on a weak foundation is the factor of safety. Typical factors of safety are 1.2 for embankments and temporary structures where the consequences of a failure are minor, to 1.5 or more for important embankments where the consequences of a failure may be catastrophic.

Design Principles

Embankment stability can be impaired if the embankment itself lacks internal stability (that is, if embankment failure occurs above foundation level) or if the foundation soil cannot adequately support the embankment (that is, if embankment failure results from excessive deformation or failure of foundation soil under the embankment load). Failure mechanisms associated with lack of "embankment internal stability" and lack of "foundation stability" are discussed in following paragraphs.

Failure Caused by Lack of Embankment Internal Stability—Although the intent of reinforcement placed at the embankment/foundation soil interface is clearly to prevent failure of the foundation soil, the effect of this layer of reinforcement on the internal stability of the embankment should not be overlooked. A layer of reinforcement at the embankment/foundation soil interface can impair the internal stability of an embankment if it acts as a slip surface, thereby causing embankment failure by lateral sliding. This mechanism is more likely to occur if the reinforcement is smooth and continuous, like some geotextiles, than discontinuous, like strips and geogrids. Even if there is no slip at the embankment/reinforcement interface, the embankment may undergo excessive lateral spreading if the reinforcement stiffness is not sufficiently large. Both embankment sliding over the reinforcement and lateral spreading of the embankment due to reinforcement elongation should be investigated during design. Embankment internal stability increases with increasing strength of the embankment fill. Thus embankment internal stability will be most critical when low strength fills are used.

For a given type and density of fill material, internal stability of an embankment is governed by the height of the embankment and the steepness of the side slopes. Both parameters also influence the magnitude and distribution of stresses on the foundation soil and, therefore, influence the stability of the foundation. In most cases, foundation soil failure is more critical than failure of the embankment itself. However, foundation soil stability and embankment internal stability are interrelated, as illustrated by the following: large deformations induced in the weak foundation soil by the embankment load may loosen and weaken the fill material, thereby impairing embankment internal stability. A fill material having cohesion may be brittle and crack at rather small strains. Cracking impairs not only the internal stability of the embankment but also the stability of the foundation soil by modifying the stress distribution at the embankment soil interface and by reducing the contribution of the fill material shear strength to the stability of the embankment/foundation system [*34*]. If necessary, embankment internal stability can be improved by layers of reinforcement placed at various levels within the embankment slopes, as discussed previously.

Failure Caused by Lack of Foundation Stability—Failure of the foundation is assumed to occur as a result of one of two mechanisms, slip surface failure and bearing capacity failure. A slip surface failure is assumed to occur when a portion of the embankment/foundation soil system slides while the other portion remains stable. The slip surface between the two portions goes through the embankment, the reinforcement, and the foundation soil. An overall bearing capacity failure is assumed to occur when the embankment as a whole punches into the foundation soil. With the first mechanism, the reinforcement breaks or is pulled from either the stable or unstable zones at its intersection with the slip surface. With the second mechanism, the reinforcement holds the embankment together; as a result, the entire embankment is assumed to act as a footing which punches into the plastic foundation soil.

Simplified Design of Embankments on Weak Foundations

In practice, it is usually assumed that the just-cited described embankment and foundation failure mechanisms can be idealized for design purposes. The three idealized failure mechanisms most often considered are [*12,35*]:

1. Lateral sliding of a portion of the embankment, which may occur along the embankment/reinforcement interface, along the reinforcement/foundation interface, or along a shallow, weak seam or layer in the foundation soil.
2. Slip surface failure through the embankment and foundation.
3. Overall bearing capacity failure of the entire embankment, which can occur only if the reinforcement is strong enough to prevent a slip surface failure.

As indicated previously, the first mechanism is related to lack of embankment internal stability while the last two mechanisms are related to lack of foundation stability. For practical design, bearing capacity is checked first because overall bearing capacity will limit embankment size and geometry and is independent of reinforcement properties. Resistance to lateral sliding is usually checked last, after reinforcement type and layout has been selected.

Lateral sliding—Lateral sliding of one portion of the embankment while the rest of the embankment remains stable can occur near the slope edge (Fig. 24*A*). The factor of safety can be determined using a general limit equilibrium wedge analysis.

An approximate analysis can be made by considering a wedge, the boundaries of which are the reinforcement and the vertical plane passing through the edge of the crest of the embankment (Fig. 24*B* and Fig. 24*C*). In this case, the driving force per unit width, P_a, is expressed by

$$P_a = 0.5 K_a \gamma H^2 \tag{13}$$

where

K_a = coefficient of active earth pressure (dimensionless),
γ = unit weight of the embankment material (kN/m^3), and
H = embankment height (m).

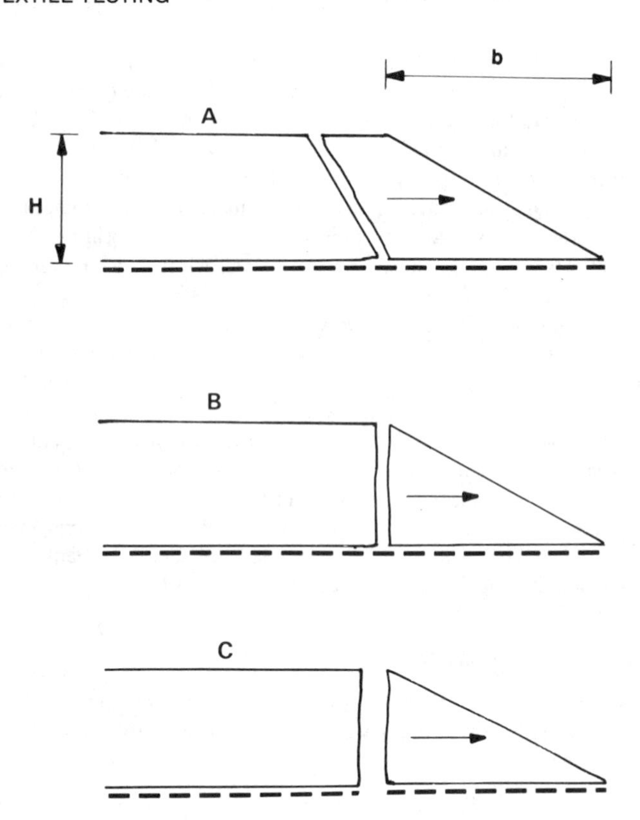

FIG. 24—*Lateral sliding of embankment:* (A) *general case;* (B) *approximate analysis when the reinforcement does not break; and* (C) *approximate analysis when the reinforcement fails.*

The resisting force, P_r, which prevents sliding of the embankment over the reinforcement, results from embankment material/reinforcement interface friction (assuming the embankment fill has zero cohesion) and is given by

$$P_r = W\lambda_1 \tan\phi = 0.5\gamma Hb\lambda_1 \tan\phi \qquad (14)$$

where

P_r = resisting force per unit width in case of sliding at the embankment/reinforcement interface (kN/m),
W = weight per unit width of the sliding portion of the embankment (kN/m),
b = width of embankment side slope (m),
λ_1 = ratio between embankment fill/reinforcement interface friction and embankment fill internal friction, and
ϕ = angle of internal friction of the embankment material. (Note that λ_1 is always less than or equal to 1.)

The factor of safety against lateral sliding of the embankment over the reinforcement is expressed by

$$FS_L = P_r/P_a = [(b/H)\lambda_1 \tan \phi]/K_a \tag{15}$$

A value of $FS_L = 2$ is typically used for design.

If FS_L is larger than 1, there will be no slip between the embankment fill and the reinforcement. In this case, the entire force per unit width, P_a, will be transmitted to the reinforcement. Part of this force is resisted by the reinforcement itself, part by reinforcement/foundation soil interface shear stresses. In the case of a frictionless foundation soil, the force per unit width resisted by the reinforcement is given by

$$\alpha = P_a - \lambda_2 b c_u \tag{16}$$

where

α = force per unit width in the reinforcement (kN/m),
P_a = driving force per unit width given by Eq 13 (kN/m),
λ_2 = ratio between reinforcement/foundation soil interface adhesion and foundation soil cohesion (undrained shear strength),
c_u = undrained shear strength of foundation soil (kN/m^2), and
b = width of embankment side slope (m). (Note that λ_2 is always less than or equal to 1.)

Development of the force given by the term $\lambda_2 c_u b$ may be uncertain in some cases, especially if the reinforcement is also performing a drainage function. Therefore, it is recommended that this term be neglected in calculating the required force per unit width in the reinforcement. Consequently, a reinforcement capable of resisting a force per unit width equal to P_a should conservatively be selected.

A certain elongation of the reinforcement is necessary to provide a force per unit width equal to P_a. To ensure that P_a is mobilized in the reinforcement before the embankment undergoes excessive lateral deformation, the reinforcement should have a high tensile stiffness. It is recommended that the tensile force per unit width, $\alpha = P_a$, be generated at a reinforcement strain of not greater than 5% if the embankment fill consists of cohesionless soil and not greater than 2% if the embankment fill is cohesive and considered to be brittle.

Slip Surface Failure—The slip surface may be assumed to be circular or noncircular, depending on the thickness of the foundation soil layer (Fig. 25). As with the case of reinforced slopes, limit equilibrium stability analyses can be used to analyze slip surface failure as long as they have been appropriately modified to account for the tensile force in the reinforcement, which is assumed to act at the reinforcement/slip surface intersection. Two approaches for incorporating this reinforcement force in slope stability analyses with circular failure surfaces were discussed previously (Fig. 16). For the case of a reinforced slope with multiple layers of reinforcement, an approach assuming that the reinforcement force modified the soil strength was suggested as being appropriate (Fig. 16*b*). The suggested technique for incorporating the reinforcement force for embankments on soft soils is different than for reinforced slopes. This is due to two factors: (1) in embankments, the component of force normal to the slip surface is usually not taken into account since its effect is localized and therefore uncertain; and (2) the reinforcement is usually at the interface of the embankment and foundation, and some portion of the reinforcement force is transmitted to the foundation soils and is not available to strengthen the embankment fill. For these reasons it is suggested that the reinforcement force be included as an independent free-body tensile force (Fig. 16*a*) for analyses of embankments over soft foundations. This approach requires selection of a reinforcement orientation and magnitude which is discussed in following paragraphs.

The orientation of the reinforcement tensile force changes as the embankment moves. Before

a

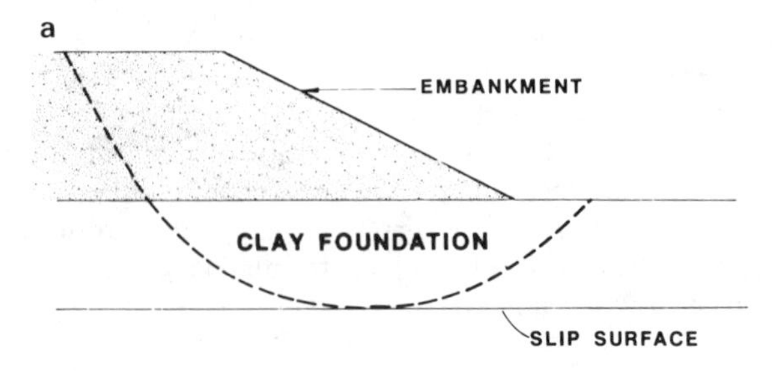

b

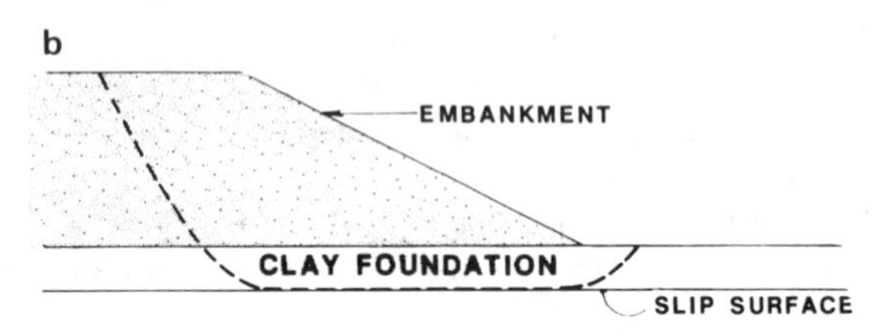

FIG. 25—*Slip surface failure:* (a) *deep-seated circular surface; and* (b) *shallow translational surface.*

any sliding movement starts, the reinforcement tensile force is horizontal. If the reinforcement is flexible and the soil stiff, and if the displacement along the slip surface is large, the reinforcement may rotate in the vicinity of the slip surface. The range of possible reinforcement orientations varies from horizontal to tangent to the slip surface. It is conservative to assume a horizontal tensile force because its contribution to the resisting moment is smaller than the contribution due to an inclined force. This conservative approach may be appropriate for embankments on brittle, sensitive foundations, embankments supporting structures, or embankments constructed with reinforcement having little flexibility. In other cases, some amount of reorientation of the reinforcement tensile force could be considered.

The magnitude of the reinforcement tensile force for use in stability analyses depends on the elongation of the reinforcement: for a given reinforcement, the larger the elongation, the larger the force. Elongation can be induced in the reinforcement as a result of:

1. Placement of reinforcement and fill.
2. Constant-volume (undrained) settlement of the foundation soil during and just after embankment construction.
3. Localized displacement associated with the development of a slip surface at the end of construction (that is, undrained failure of the foundation soil) or after a period of undrained creep.
4. Settlement due to consolidation of the foundation soil (that is, settlement of the foundation soil in the drained state).

These four causes of reinforcement elongation are discussed as follows:

1. It may be possible to induce some elongation in the reinforcement by laying it taut and by adopting an appropriate sequence of fill placement. The importance of proper construction procedures for reinforced embankments, especially on very soft foundations, cannot be overem-

phasized. Low ground pressure equipment, shallow initial lift thicknesses, partially loaded haul vehicles, and careful site preparation are all steps that can be taken to improve construction. Fill placement, spreading, and compaction procedures are also important. Christopher and Holtz [*12*] and Holtz [*36*] review fill placement procedures designed to maximize reinforcement tension and construction stability. For high tensile stiffness reinforcement, these procedures should not induce more than 1 to 2% strain in the reinforcement.

2. The foundation soil deforms and settles during embankment construction as a result of the increasing applied loads. Lateral movements of the foundation soil in the vicinity of the embankment/foundation soil interface tend to stretch the reinforcement. As indicated previously, lateral movements may be large with low permeability soils such as clays, and may be smaller with high permeability soils such as peats. The magnitude of the lateral movements also depends on the thickness of the foundation soil layer. Large lateral movements are more likely to occur with relatively shallow layers since vertical movements are constrained. Nonuniform foundation thickness will also induce lateral movements. Rowe and Soderman [*37*] conducted a finite-element study to determine lateral movements at the embankment/foundation soil interface. The results of this study indicated tensile strains induced in the reinforcement as a result of settlement during construction on the order of 1 to 9% depending on embankment height, foundation soil properties, and soil/reinforcement interaction.

3. Prior to failure, strains in the reinforcement are equal to the horizontal strains in the adjacent soil (assuming no slip). Once a slip surface failure begins, additional elongation is induced in the reinforcement at the same time as large deformations appear at the crest of the embankment. This additional elongation is localized in a section of reinforcement located in the vicinity of the slip surface and need not be compatible with strains in the adjacent soil. The additional reinforcement tensile force generated by this elongation may or may not restore stability depending on the stiffness of the reinforcement and on the strain-softening characteristics of the foundation. Since most foundations are at least slightly strain softening, the reinforcement tensile force increase associated with this increment of strain should probably be neglected.

4. Additional settlement will progressively take place with time as a result of foundation soil consolidation (that is, progressive expulsion of water from soil). This settlement, resulting from a decrease in volume of the soil, is not normally associated with significant lateral movement and, consequently, is not expected to induce much additional elongation in the reinforcement.

Selection of a strain level for evaluation of tensile resistance for design using the ASTM wide strip test method (D 4595-86) should take into account the four causes of reinforcement strain just described. In most practical problems, strains induced by soil consolidation can be disregarded because soil consolidation does not cause large lateral movement and will increase soil strength, thereby progressively decreasing the need for tensile reinforcement. For soils that consolidate and gain strength with time, the critical design condition can therefore be assumed to be at the end of construction.

The selected strain level should also consider the potential for progressive failure of the foundation due to excessive deformation and strain softening. The reinforcement should provide adequate reinforcing force, should a slip surface begin to develop, to prevent strain softening and progressive failure. For highly sensitive, brittle clay foundations, a limiting strain in the range of 2 to 3% may be appropriate. Limiting strains on the order of 4 to 6% may be acceptable for medium to low sensitivity clay foundations. Finally, if the foundation soils are not sensitive (this category may include peats, marsh deposits, and young normally consolidated clays), and if some differential settlement and/or spreading of the embankment crest can be tolerated, limiting reinforcement strains up to as large as 10% may be appropriate.

The stability of an embankment and foundation with respect to a slip surface failure can be evaluated using classical limit equilibrium procedures incorporating a reinforcement tensile force determined as just described. Charts presenting the results of limit equilibrium slip surface analyses have been published by several authors [*38–42*]. The authors have found the charts prepared by Milligan and Busbridge (one is shown in Fig. 26) to be conservative and useful. The charts were developed based on (1) moment equilibrium along the critical circular arc through the foundation and Coulomb wedge through the embankment, and (2) horizontal force equilibrium along a critical multipart wedge. The latter equilibrium condition was found to control for foundation depth to embankment height ratios of less than about 0.5. These charts have been found to be conservative and are included herein as a simple means to obtain a preliminary indication of the influence of tensile reinforcement on a given design.

Bearing Capacity Failure—Bearing capacity failure may occur if the reinforcement does not break and holds the embankment together. The embankment then acts as a foundation which may punch into the foundation soil. Classical methods for the bearing capacity of shallow foundations can be used.

Bearing capacity failure does not occur if the maximum stress exerted by the embankment on the foundation soil is smaller than the foundation soil bearing capacity. This is expressed by the following equation:

$$\gamma H < c_u N_c \tag{17}$$

where

γH = maximum stress exerted by the embankment on the foundation soil (kN/m^2),
γ = unit weight of soil (kN/m^3),
H = embankment height, m,
$c_u N_c$ = bearing capacity of the foundation soil (kN/m^2),
c_u = undrained shear strength (cohesion) of the foundation soil (kN/m^2), and
N_c = bearing capacity coefficient (dimensionless).

The bearing capacity coefficient N_c is given in Fig. 27 as a function of the ratio between the "average width" of the embankment, B, and the thickness, D, of the foundation soil layer.

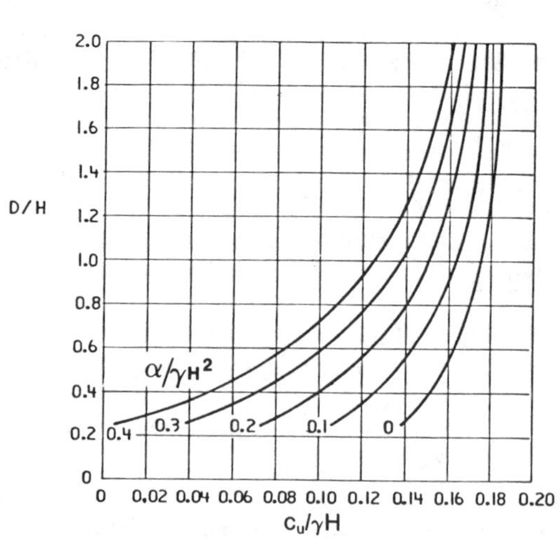

FIG. 26—*Design of embankment on weak foundation: Chart giving the required force per unit width, α, in the reinforcement as a function of: γ = unit weight of fill material; H = height of embankment; c_u = undrained shear strength of the foundation soil; and D = depth of foundation soil. The side slopes are 2:1 (horizontal: vertical) and the friction angle of the embankment fill is 30°. (After Milligan and Busbridge* [39].*)*

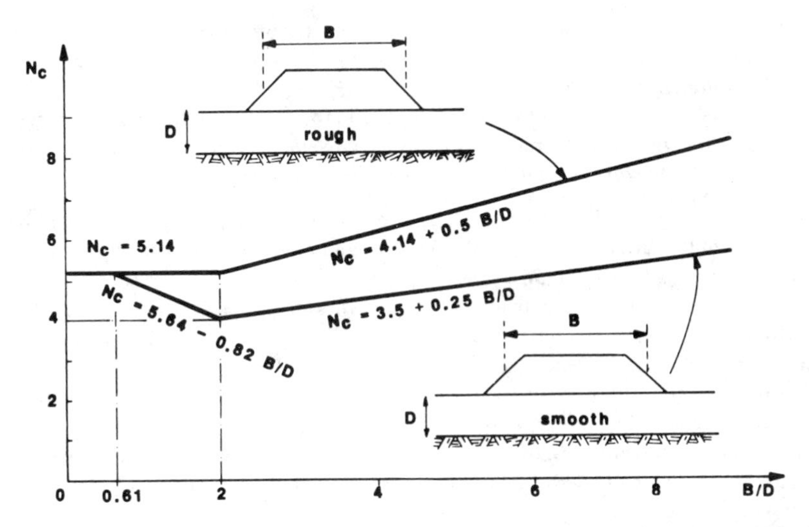

FIG. 27—*Design of embankment of weak foundation: bearing capacity coefficient* N_c *(after Mandel and Salencon* [43,44], *and Giroud et al.* [45].*)*

As indicated by Christopher and Holtz [*12*], other equations have been proposed to verify the bearing capacity of a finite layer of foundation soil under an embankment. These include:

$$\gamma H < 2c_u B/D \tag{18}$$

and

$$\gamma H < 4c_u \tag{19}$$

Comparing these two equations with Eq 17 and Fig. 27, it appears that:

1. Equation 18 gives bearing capacity values that may not be conservative when B/D is larger than 3. Therefore, use of Eq 18 is not recommended for these cases.
2. Equation 19 corresponds to the lower bound shown in Fig. 27. Use of Eq 19 is overconservative when B/D is large (for example, larger than 5).

Reinforcement Properties and Relevant Test Methods

The reinforcement properties and relevant test methods for design of embankments over weak foundations are similar to those described previously for reinforced slopes. Evaluation of the tensile resistance for design should be based on the ASTM proposed wide strip tension test. The evaluation should be carried out using the creep modification to the wide strip test when feasible. When creep testing is not feasible, evaluation of tensile resistance for design may be carried out using the wide strip test at the standard rate of strain. The results from this test should be modified to reflect the influence of creep. While the duration for which tensile reinforcement is needed is often limited (due to foundation strength gain associated with soil consolidation), the influence of creep can usually not be neglected. This is due to the fact that a major portion of a polymer's creep strength loss occurs within the first several months of loading. The isochronous load-strain curve for design should correspond to the appropriate load duration. Long-term in-ground deterioration may be insignificant, however, for applications in which the foundation soils gain strength with time. Whether creep testing or constant rate-of-strain testing is carried out, the tensile resistance for design should be based on limiting strains not larger than those recommended previously.

For the case of an embankment over a weak foundation only one layer of reinforcement is typically used, while multiple layers are normally used in slope reinforcement applications. This factor leads to the following special considerations for embankments over weak foundations:

1. Since only one layer of reinforcement is used, material homogeneity and strict manufacturing quality control are essential. Quality control and continuous construction inspection of sewn geotextile seams is critical [*36*].
2. During the first stage of construction (while a working pad is being established), foundation deformations under construction traffic wheel loads may cause reinforcement damage. Assessment of the potential for construction site damage may be important.
3. To prevent internal failure of the embankment, high reinforcement tensile stiffness and high soil-reinforcement friction/interlocking is desirable.
4. The pullout behavior of reinforcement is usually not critical for cohesionless embankments constructed over soft clays. Soil reinforcement interface friction behavior should be modelled using direct shear tests.
5. Reinforcement forces and strains can be developed parallel to the embankment centerline in certain cases. The potential for longitudinal reinforcement forces and strains may occur: (1) during construction over very weak sites prone to mud waving; (2) at the ends of an embankment; and (3) due to differential settlements and bending of embankments built over nonuniform foundation conditions. When the just-cited conditions prevail, longitudinal reinforcement forces and strains should be considered during design.
6. Seam strengths between adjacent layers of reinforcement must transmit the reinforcement forces, both during the initial construction stages and long term. Adequate seam strength is therefore essential. Christopher and Holtz [*12*] should be referenced for a detailed discussion of seam strength.

Acknowledgments

This paper was written while the senior author was on the staff of The Tensar Corp., Morrow, GA. The authors would like to thank J. R. Busbridge, R. K. Rowe, and R. A. Jewell for the comments each made on the development and use of their respective design charts. The authors also thank T. S. Ingold for providing them with a copy of his yet unpublished paper and design charts. Special thanks are due to Doris Campbell and Barbara Hutcheson, who expertly typed the paper, and Rick Komada, who drafted the figures.

References

[*1*] Leflaive, E., "Soil Reinforced with Continuous Yarns: The Texsol," *Proceedings of the Eleventh International Conference on Soil Mechanics and Foundation Engineering*, San Francisco, Vol. 3, 1985, pp. 1787-1790.

[*2*] McGown, A., Andrawes, K. Z., Hytiris, N., and Mercer, F. B., "Soil Strengthening Using Randomly Distributed Mesh Elements," *Proceedings of the Eleventh International Conference on Soil Mechanics and Foundation Engineering*, San Francisco, Vol. 3, 1985, pp. 1735-1738.

[*3*] Hausmann, M. R. and Vagneron, J. M., "Analysis of Soil Fabric Interaction," *Proceedings of the International Conference on the Use of Fabrics in Geotechniques*, Paris, Ecole Nationale des Ponts et Chaussées, Paris, Vol. 3, 1977, pp. 139-144.

[*4*] Koerner, R. M. and Welsh, J. P., *Construction and Geotechnical Engineering Using Synthetic Fabrics*, Wiley-Interscience, New York, 1980.

[5] Rankilor, P. R., *Membranes in Ground Engineering,* Wiley, New York, 1981.
[6] Giroud, J. P. and Carroll, R. G., "Geotextile Products," *Geotechnical Fabric Reports,* Vol. 1, No. 1, 1983, pp. 12-15.
[7] Giroud, J. P., Arman, A., and Bell, J. R., "Geotextiles in Geotechnical Engineering Research and Practice," report of the ISSMFE Technical Committee on Geotextiles, *Geotextiles and Geomembranes,* Vol. 2, No. 3, 1985, pp. 174-242.
[8] McGown, A., Andrawes, K. Z., Yeo, K. C., and DuBois, D. D., "The Load-Strain-Time Behavior of Tensar Geogrids," *Proceedings of the Symposium on Polymer Grid Reinforcement in Civil Engineering,* 1984*a,* The Institution of Civil Engineers, London, pp. 11-17.
[9] McGown, A., Paine, N., and DuBois, D. D., "Use of Geogrid Properties in Limit Equilibrium Analysis," *Proceedings of the Symposium on Polymer Grid Reinforcement in Civil Engineering,* 1984*b,* The Institution of Civil Engineers, London, pp. 31-35.
[10] Bell, J. R. and Hicks, R. G., et al., "Evaluation of Test Methods and Use Criteria for Geotechnical Fabrics in Highway Applications," Report No. FHWA/RD-80/021, Oregon State University, Corvallis, OR, 1980.
[11] McGown, A., Andrawes, K. Z., and Kabir, M. H., "Load Extension Testing of Geotextiles Confined in Soils," *Proceedings of the Second International Conference on Geotextiles,* Las Vegas, Vol. 3, Industrial Fabrics Association International, St. Paul, MN, 1985, pp. 793-798.
[12] Christopher, B. R. and Holtz, R. D., *"Geotextile Engineering Manual,"* Federal Highway Administration, Washington, DC, 1985.
[13] Ward, I. M., "The Orientation of Polymers to Produce High Performance Materials," *Proceedings of the Symposium on Polymer Grid Reinforcement in Civil Engineering,* The Institution of Civil Engineers, London, 1984, pp. 4-10.
[14] Collios, A., Delmas, P., Gourc, J. P., and Giroud, J. P., "Experiments on Soil Reinforcement with Geotextiles," *The Use of Geotextiles for Soil Improvement,* Preprint 80-177, American Society of Civil Engineering, Portland, pp. 53-73.
[15] Haliburton, T. A., Anglin, C. C., and Lawmaster, J. D., "Testing of Geotechnical Fabric for Use as Reinforcement," *Geotechnical Testing Journal,* Vol. 1, No. 4, American Society for Testing and Materials, Philadelphia, 1978, pp. 203-212.
[16] Holtz, R. D., "Laboratory Studies of Reinforced Earth Using a Woven Polyester Fabric," *Proceedings on the International Conference on the Use of Fabrics in Geotechniques,* Paris, Vol. 3, Association Amicale des Ingénieurs, Paris, 1977, pp. 149-154.
[17] Ingold, T. S. and Templeman, J. E., "The Comparative Performance of Polymer Net Reinforcement," *Proceedings of the International Conference on Soil Reinforcement,* Paris, Vol. I, Ecole Nationale des Ponts et Chaussées, Paris, 1979, pp. 65-70.
[18] Romstad, K. M., Herrmann, L. R., and Shen, C. K., "Pull-out Testing of Tensar SR2 Geogrids, Final Report for The Tensar Corporation," University of California, Davis, 1985.
[19] McGown, A., Discussion to Session 8, *Proceedings of the VII European Conference on Soil Mechanics and Foundation Engineering,* Brighton, England, Vol. 4, International Society of Soil Mechanics and Foundation Engineering, U. of California, Berkeley, CA, pp. 285-286.
[20] Andrawes, K. Z., McGown, A., Mashhour, M. M., and Wilson-Fahmy, R. F., "Tension Resistant Inclusions in Soils," *Journal of the Geotechnical Engineering Division,* Vol. 106, No. GT12, American Society of Civil Engineers, New York, 1980, pp. 1313-1326.
[21] Murray, R. T., "Fabric Reinforcement of Embankments and Cuttings," *Proceedings of the Second International Conference on Geotextiles,* Las Vegas, Vol. 3, Industrial Fabrics Association International, St. Paul, MN, 1982, pp. 707-713.
[22] Jewell, R. A., Paine, N., and Woods, R. I., "Design Methods for Steep Reinforced Embankments," *Proceedings of the Symposium on Polymer Grid Reinforcement in Civil Engineering,* The Institution of Civil Engineers, London, 1984, pp. 70-81.
[23] Schneider, H. R. and Holtz, R. D., "Design of Slopes Reinforced with Geotextiles and Geogrids," *Geotextiles and Geomembranes,* Vol. 4, No. 2, 1986.
[24] Schmertmann, G. R., Bonaparte, R., Chouery, V. C., and Johnson, R. J., "Design Charts for Geogrid Reinforced Soil Slopes," *Proceedings of Geosynthetics '87,* New Orleans, Industrial Fabrics Association International, St. Paul, 1987, pp. 108-120.
[25] Bonaparte, R., Schmertmann, G. R., and Williams, N. D., "Seismic Design of Slopes Reinforced with Geogrids and Geotextiles," *Proceedings of the Third International Conference on Geotextiles,* Vienna, Vol. 1, 1986, pp. 273-278.
[26] Jones, C. J. F. P., *Earth Reinforcement and Soil Structures,* Butterworths Advanced Series in Geotechnical Engineering, London, 1985.
[27] Schlosser, F., "La terre armee', historique development actuel et futur," *Proceedings of the Symposium on Soil Reinforcing and Stabilizing Techniques,* NWSIT/NSW University, 1978, pp. 5-28.

[28] Steward, J., Williamson, R., and Mohney, J., "Guidelines for Use of Fabrics in Construction and Maintenance of Low-Volume Roads," U.S. Department of Agriculture, Forest Service, Portland, OR, 1977 (also published as Report No. FHWA-TS-78-205 by the Federal Highway Administration, Washington, DC).
[29] Broms, B. B., "Design of Fabric Reinforced Retaining Structures," *Proceedings, Symposium on Earth Reinforcement,* American Society of Civil Engineers, Pittsburgh, 1978, pp. 282-304.
[30] Berg, R. R., Bonaparte, R., Anderson, R. A., and Chouery, V. E., "Design, Construction and Performance of Two Geogrid Reinforced Soil Retaining Walls," *Proceedings of the Third International Conference on Geotextiles,* Vienna, Vol. 2, 1986, pp. 401-406.
[31] Ingold, T. S., *Reinforced Earth,* Thomas Telford, Ltd, London, 1982.
[32] Poulos, H. G. and Davis, E. H., *Elastic Solutions for Soil and Rock Mechanics,* John Wiley and Sons, New York, 1974.
[33] U.S. Navy, *Soil Mechanics,* Design Manual (DM) 7.1, and *Foundations and Earth Structures,* Design Manual (DM) 7.2, Naval Facilities Engineering Command, U.S. Government Printing Office, Washington, DC, 1982.
[34] Chirapuntu, S. and Duncan, J. M., "The Role of Fill Strength in the Stability of Embankments on Soft Clay Foundations," Geotechnical Engineering Research Report, Department of Civil Engineering, University of California, Berkeley, 1975.
[35] Jewell, R. A., "A Limit Equilibrium Design Method for Reinforced Embankments on Soft Foundations," *Proceedings of the Second International Conference on Geotextiles,* Las Vegas, Vol. 1, Industrial Fabrics Association International, St. Paul, 1982, pp. 671-676.
[36] Holtz, R. D., "Soil Reinforcement with Geotextiles," *Third NTI International Geotechnical Seminar,* Nanyang Technological Institute, Singapore, 1985, pp. 55-74.
[37] Rowe, R. K. and Soderman, K. L., "An Approximate Method for Estimating the Stability of Geotextile Reinforced Embankments," *Canadian Geotechnical Journal,* Vol. 22, No. 3, 1985.
[38] Rowe, R. K., "Reinforced Embankments: Analysis and Design," *Journal of Geotechnical Engineering Division,* Vol. 110, No. GT2, American Society of Civil Engineers, New York, 1984, pp. 231-246.
[39] Milligan, V. and Busbridge, J. R., "Guidelines for the Use of Tensar in Reinforcement of Fills Over Weak Foundations," Golder Associates report to The Tensar Corp., Mississauga, Ontario, 1983, to be published.
[40] Gourc, J. P., *"Quelques aspects du comportement des geotextiles en mecanique des sols,"* Ph.D. thesis submitted at the University of Grenoble, Grenoble, France, 1982.
[41] Fowler, J., "Theoretical Design Considerations for Fabric-Reinforced Embankments," *Proceedings of the Second International Conference on Geotextiles,* Las Vegas, Vol. 4C, Industrial Fabrics Association International, St. Paul, MN, 1982, pp. 665-670.
[42] Ingold, T. S., "Analysis of Geotextile Reinforced Embankments Over Soft Clays," to be published.
[43] Mandel, J. and Salencon, J., "Force portante d'un sol sur une assise rigide," *Proceedings of the Seventh International Conference on Soil Mechanics and Foundation Engineering,* Mexico, Vol. 2, 1969, pp. 157-164.
[44] Mandel, J. and Salencon, J., "Force portante d'un sol sur assise rigide—Etude theorique," *Geotechnique,* Vol. 22, No. 1, 1972, pp. 79-93.
[45] Giroud, J. P., Nhiem, T. V., and Obin, J. P., *"Tables pour le calcul des foundations,"* Vol. 3, Dunod, Paris, 1973.

Durability of Geotextiles

Jack Hodge[1]

Durability Testing

REFERENCE: Hodge, J., "**Durability Testing,**" *Geotextile Testing and the Design Engineer, ASTM STP 952,* J. E. Fluet, Jr., Ed., American Society for Testing and Materials, Philadelphia, 1987, pp. 119-121.

ABSTRACT: Durability, while an essential requirement for geotextiles, is difficult to predict through testing. Since applications are so varied and field conditions so diverse, it is a formidable task to provide test data that can help the design engineer choose the correct engineering fabric (construction, weight, physical properties, etc.) for the intended application. Virtually all durability tests are "index" tests which allow the design engineer to compare the performance of different engineering fabrics, tested in an identical manner, and from these fabrics to select the geotextile that best meets his requirements.

KEY WORDS: durability, geotextile, index test

Durability Testing

Durability, while an essential requirement for geotextiles, is a quality that is difficult to predict by laboratory testing. Field applications are so diverse and conditions so varied that it is a formidable task to provide test data that can help the engineer choose the correct geotextile (construction, weight, physical properties, etc.) for the intended application. Virtually all durability tests are "index" tests which allow the design engineer to compare the performance of different engineering fabrics, hopefully tested by the same procedures, and from these fabrics to select the one that best meets his requirements.

While durability criteria can include a host of physical properties such as resistance to puncture, tear, cutting, etc., most criteria are considered as mechanical properties of the geotextile. Among those considered endurance properties and which will be discussed in this paper are:

1. Ultraviolet light stability.
2. Abrasion resistance.
3. Chemical stability.
4. Thermal stability.

Ultraviolet Light Stability

Geotextiles are generally protected from ultraviolet (UV) degradation by either the chemical makeup of the polymer or by the addition of additives such as carbon black. In addition, most geotextile applications involve covering the fabric with soil, thereby reducing the problem of degradation from UV radiation. However, geotextiles are used in applications such as silt fences, erosion control, and retaining walls where the effect of exposure to sunlight on the geotextile could be a factor. Also, geotextiles are often stored on the job site for varying lengths of time, and this type of exposure must be considered.

A test method has been adopted by ASTM Committee D35 on Geotextiles, Geomembranes,

[1] Manager, Quality Control Spunbond, Hoechst Fibers Industries, Spartanburg, SC 29304.

and Related Products to provide a UV resistance measurement value that could be used to compare fabrics. This method [ASTM Test Method for Deterioration of Geotextile from Exposure to Ultraviolet Light and Water (Xenon-Arc Type Apparatus) (D 4355-84)] calls for the exposure of geotextile fabric specimens for 0, 150, 300, and 500 h of UV exposure in a xenon-arc light source device. The exposure consists of 120-min cycles as follows: 102 min of light only, followed by 18 min of water spray and light. The effect of exposure is determined by comparing the strip tensile strength of the unexposed (0 h) material and the tested material.

In addition to the test method discussed above, Committee D35 also has a task group considering guidelines for the on-site protection and handling of geotextiles.

Abrasion Resistance

Abrasion to a geotextile can come from friction produced by various types of movement of rock and/or soil against the surface of the fabric such as wave action or riprap, sand scour sediment in a stream, or the aggregate cover in fabric reinforcement applications such as roadways and railroads. Abrasion also often comes during installation from equipment and personnel.

The tabor test [ASTM Test Method for Abrasion Resistance of Textile Fabrics (Rotary Platform, Double-Head Method)], in which the fabric is subjected to the rotary action of various abrasive wheels, was adapted from the textile industry and is found in many specifications. In this method the abrasion is measured by the number of revolutions required to wear through the specimen or the percent strength retained, as measured by a strip tension test. Several factors in this method have caused producers and users to consider it unacceptable for geotextiles, mainly the rotary type of abrasion which is generally unrelated to geotechnical applications and the small specimen size (1 in.) which produces variable results.

The ASTM task group on Abrasion is currently investigating a new abrasion test for geotextiles in which the abrasion is produced by a reciprocal back-and-forth rubbing motion of a sandpaper abradant against the geotextile. The instrument used in this test is a Stoll Flex abrader modified to allow the fabric to be mounted on a stationary platform and the abrading medium on a reciprocating platform. The abrader can be loaded to provide a constant pressure and has an adjustable speed drive to allow the abrasive action to be controlled as required. In addition to allowing flexibility in the applied load, abrasive medium, and speed of the abrading cycles, the test area allows a 76.2-mm (3-in.)-wide area to be abraded and a 50.8-mm (2-in.)-wide strip tension test to be used to measure strength loss in the fabric. A round-robin test series is currently underway to provide more information concerning the data provided by the test regarding abrasive effect, reproducibility, etc. While more information is needed, the results from preliminary testing in different laboratories are encouraging.

Chemical Stability

While most geotextiles are resistant to deterioration from the normal chemical environment found in soils because of their chemical structure, unusual environments such as hazardous waste disposal sites may require special evaluation.

There is no current established test for measuring the chemical stability of geotextiles. The ASTM task group on Chemical Stability currently has a proposed method being balloted at the subcommittee level concerning the resistance of geotextiles to chemical environments. In this method the geotextile is immersed in either a selected chemical reagent for a specific application or a series of standard chemical reagents. The exposure time recommended is four weeks at room temperature. After exposure, the tensile properties are measured using the 50.8-mm (2-in.)-strip tension test {ASTM Test Methods for Breaking Load and Elongation of Textile Fabrics [D 1682-64 (1975)]} and compared to the tensile strength of unexposed specimens.

Thermal Stability

All the current geotextile polymers are relatively stable under normal temperature ranges. However, certain changes in mechanical properties may occur in some instances, especially in relation to stress-strain characteristics. A test method is currently under development by the ASTM D35 task group on Thermal stability. The method covers a procedure for determining the effects of climatic temperature on the tensile strength and elongation properties of geotextiles.

The procedure involves conditioning geotextile specimens at selected temperatures in an environmental chamber attached to a tension testing machine, then testing specimens using strip tension tests per ASTM D 1682. The test results are compared to those obtained on specimens conditioned and tested under standard laboratory conditions.

Conclusion

While durability is a difficult property to define by means of laboratory testing, index testing does allow comparison of different engineering fabrics. Test methods on UV stability and thermal stability have already been adopted by ASTM Committee D35, and other standard practices involving such characteristics as abrasion resistance, chemical stability, and creep resistance are currently under development. These methods will provide the design engineer with information that can be used in the selection of the proper engineering fabric for the desired application.

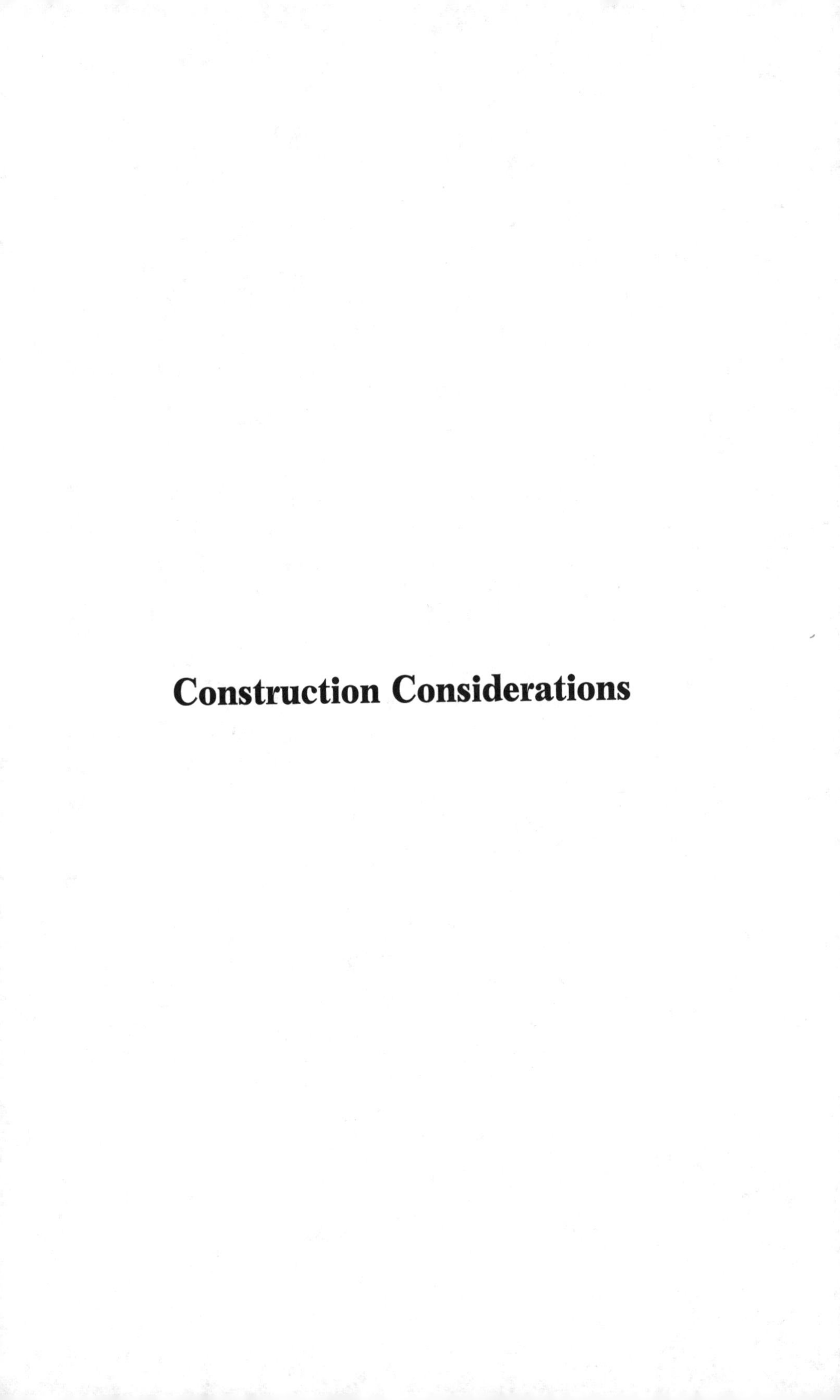

Construction Considerations

Gregory N. Richardson[1] *and David C. Wyant*[2]

Geotextiles Construction Criteria

REFERENCE: Richardson, G. N. and Wyant, D. C., **"Geotextiles Construction Criteria,"** *Geotextile Testing and the Design Engineer, ASTM STP 952,* J. E. Fluet, Jr., Ed., American Society for Testing and Materials, Philadelphia, 1987, pp. 125–138.

ABSTRACT: Designers of geotextile systems must place greater emphasis on a review of construction criteria compared to typical civil engineering design. This enhanced awareness of the construction procedure and relevant criteria is necessitated by the lack of significant construction experience in the use of these materials by most contractors and field personnel. This paper reviews the four primary roles of geotextiles and discusses specific construction criteria for each. Information presented in this paper has been obtained primarily from two editions of the Geotextile Engineering Manual, as compiled for the Federal Highway Administration, and from actual field and contractual experience of the authors.

KEY WORDS: geotextile, drainage, erosion, separation, reinforcement, survivability

Civil engineering designers in today's practice are rarely concerned with specific construction details associated with the implementation of their designs. Specifications are performance- and not construction-oriented. Thus, a structural engineer does not typically review or specify construction procedures to be used in the erection of a particular building. Tradition and legal precedent have worked to make the contractor responsible for construction procedures. A successful contractor is one who understands the general mechanics of the structure and uses this knowledge or experience to efficiently sequence the construction. The use of geotextiles is, however, in such an early stage of development that few contractors, and many designers, do not understand the roles or capacities of geotextiles in systems under construction. When this condition is anticipated, the design engineer must provide clear construction criteria that will minimize field problems. It should be noted, however, that these criteria may be construed as enlarging the legal liability of the designer.

Successful construction criteria ensure that the geotextile will survive installation and that the construction sequencing will not lead to failure of the existing subgrade. As such, this criteria may be dependent on fabric properties and upon the role that the geotextile is intended to perform in the completed structure. This paper reviews common installation criteria associated with fabric survivability and the four major roles played by geotextiles: (1) drainage; (2) erosion control; (3) separation; and (4) reinforcement. The geotextile will play several of these roles in most design, so that the construction criteria must reflect all roles.

Construction-related problems common to the use of geotextiles can generally be associated with the following conditions:

1. Field fill placement or compaction techniques damage the geotextile.
2. Installation loads are significantly in excess of design loads, leading to failure during construction.

[1] Corporate engineer, Soil & Material Engineers, Inc., Cary, NC 27511.
[2] Research scientist, Virginia Transportation Research Council, Charlottesville, VA 22903.

3. Construction environment leads to a significant reduction in assumed fabric properties, causing a failure of the completed project.
4. Field seaming or overlap of the geotextile fails to fully develop desired fabric mechanical properties.
5. Instabilities during various construction phases may render a design inadequate even though the final profile would have been stable.

This paper reviews both general geotextile and construction criteria developed to minimize field installation problems. Field inspection by a qualified engineer must be considered vital to the successful installation in all performance modes.

Geotextile Survivability

The ability of a geotextile to survive installation is dependent upon both the physical properties of the fabric and the direct environment it is placed in. Important properties of a geotextile based on constructibility considerations are listed in Table 1 for the four primary role functions [*1*]. Existing textile-related ASTM tests for these properties are presented when possible, although the reader should be cautioned that all tests are currently under revision by ASTM D-35[3] for specific application to geotextiles. In general, the geotextile must have properties that enable its design function to be performed and at the same time must not create unusual field handling problems. A typical field handling problem is generated by exceptionally large rolls of fabric that have been exposed to rain and have absorbed a significant quantity of water, posing

TABLE 1—*Important geotextile properties—constructability.*

Physical Property	Role				Test Procedure
	Drainage	Erosion	Separation	Reinforcement	
1. Adsorption	Yes	Yes	Yes	Yes	None
2. Cutting resistance	Yes	Yes	Yes	Yes	None
3. Flammability	...	Yes	...	...	None
4. Flexibility	Yes	Yes	Yes	...	ASTM D 1388[a]
5. Modulus	...	...	Yes	Yes	Proposed by ASTM Committee D-35
6. Puncture resistance	Yes	Yes	Yes	Yes	ASTM D 751[b]
7. Roll dimensions	Yes	Yes	Yes	Yes	N/A
8. Seam strength	...	Yes	...	Yes	ASTM D 1682, Method G[c]
9. Specific gravity	...	Yes	Yes	...	ASTM D 854[d]
10. Tear strength	Yes	Yes	Yes	Yes	ASTM D 1117[e]
11. Tensile strength	Yes	Yes	Yes	Yes	ASTM D 1682, Method G
12. Temperature stability	...	...	...	Yes	None
13. UV stability	Yes	Yes	Yes	Yes	ASTM D 4355[f]
14. Weight	Yes	Yes	Yes	Yes	N/A
15. Wet and dry stability	...	...	...	Yes	None

[a]ASTM Test Methods for Stiffness of Fabrics [D 1388-64 (1975)].
[b]ASTM Method of Testing Coated Fabrics (D 751-79).
[c]ASTM Test Methods for Breaking Load and Elongation of Textile Fabrics [D 1682-64 (1975)].
[d]ASTM Test Method for Specific Gravity of Soils (D 854-83).
[e]ASTM Methods of Testing Nonwoven Fabrics (D 1117-80).
[f]ASTM Test Method for Deterioration of Geotextiles from Exposure to Ultraviolet Light and Water (Xenon-Arc Type Apparatus) (D 4355-84).

[3]ASTM Committee D-35 on Geotextiles, Geomembranes, and Related Products.

a major problem to the contractor to simply move and unroll. The geotextile in this case may be damaged by its own weight.

Construction Criteria—Drainage

Field applications using geotextiles in a drainage role are those where the flow of water is primarily in the plane of the fabric. These applications include both vertical and base drains in retaining walls and gas or water drains below membranes or other impervious layers. Drainage layers may play multiple roles in the design. Horizontal drainage layers may provide reinforcement if large strains are anticipated and filtration if the anticipated flow of water includes a sizable component normal to the geotextile. When a multiple role is required, then the construction criteria for both roles should be included in the project specifications.

Assuming that an appropriate geotextile has been selected, the next challenge to its survivability is placing the first layer of fill upon it. This contact layer of fill may consist of sand, gravel, or even 227 kg (500 lb) riprap. The survivability of the fabric is dependent upon the condition of the existing subgrade that it covers, the nature of the material placed upon it, and the manner in which the material is placed. The degree of fabric strength required to ensure survivability as a function of subgrade conditions, construction, and cover material is shown in Table 2 [*2*]. Recommended minimum physical geotextile properties for the four major roles are given in Table 3 [*3*]. The definition of protected and unprotected fabrics is role dependent and is defined in Table 3.

In a purely drainage mode, the geotextile's function is not impaired by reasonable penetrations through the fabric. The in-plane flow of water can simply flow around any intrusions without causing a significant decrease in the drainage capacity. A drain formed using a geotextile can be damaged during construction only if the requisite collector drains, weep holes, etc. are either faulty in design or construction.

Composite drainage systems consist of a high porosity layer such as a mat sandwiched between one or two layers of geotextile. This composite system offers significantly higher values of transmissibility than geotextiles alone but may be collapsed if the normal pressures acting on the system become excessive. An example of this concern is the use of a composite drainage system on the back face of a retaining wall. If excessive compaction forces are used in the compaction of fill behind the wall, then the composite system could be collapsed and fail to function as a drainage layer. It is easy to see that a similar failure could occur on a horizontal drainage layer due to excessive vehicle loading during construction.

Construction Criteria—Erosion Control

Erosion control is an application of the geotextile filtration role where the flow of water may be either normal or tangential to the surface of the geotextile. Unlike drainage, erosion control is concerned with the flow of water normal to the fabric and not within its plane. As such, erosion control applications are more sensitive to construction-generated penetrations or tears. Erosion control applications include trench drains, silt fences, and reverse filters such as used in embankment protection and surface erosion control.

Construction specifications for trench drains and associated systems should include the following criteria:

1. Trench sides and base should be excavated to provide a smooth and fairly level surface. Zones of the sidewall that have collapsed should be cleaned and all major depressions filled with granular material so that the geotextile will not be distorted or torn.
2. Care should be taken to place the fabric in intimate contact with the soil so that no void

TABLE 2—*Required degree of fabric survivability as a function of subgrade conditions and construction equipment.*

	Construction Equipment and 15-30 cm. (6-12 in.) Cover Material Initial Lift Thickness		
Subgrade Conditions	Low Ground Pressure Equipment, $\leq$0.28 kg/cm^2 ($\leq$4 psi)	Medium Ground Pressure Equipment, $>$0.28 kg/cm^2, $\leq$0.56 kg/cm^2 ($>$4 psi, $\leq$8 psi)	High Ground Pressure Equipment, $>$0.56 kg/cm^2 ($>$8 psi)
Subgrade has been cleared of all obstacles except grass, weeds, leaves, and fine wood debris. Surface is smooth and level such that any shallow depressions and humps do not exceed 15 cm (6 in.) in depth and height. All larger depressions are filled. Alternatively, a smooth working table may be placed.	Low	Moderate	High
Subgrade has been cleared of obstacles larger than small- to moderate-sized tree limbs and rocks. Tree trunks and stumps should be removed or covered with a partial working table. Depressions and humps should not exceed 45 cm (18 in.) in depth and height. Larger depressions should be filled.	Moderate	High	Very high
Minimal site preparation is required. Trees may be felled, delimbed, and left in place. Stumps should be cut to protect not more than 15 cm (6 in.) above subgrade. Fabric may be draped directly over the tree trunks, stumps, large depressions, and humps, holes, stream channels, and large boulders. Items should be removed only if placing the fabric and cover material over them will distort the finished road surface.	High	Very high	Not recommended

NOTE:

1. Recommendations are for 15 to 30 cm (6 to 12 in.) initial lift thickness. For other initial lift thicknesses:

30 to 45 cm (12 to 18 in.):	Reduce survivability requirement 1 level
45 to 60 cm (18 to 24 in.):	Reduce survivability requirement 2 levels
$>$60 cm ($>$24 in.):	Reduce survivability requirement 3 levels

Survivability levels are, in increasing order: low, moderate, high, and very high.

2. For special construction techniques such as prerutting, increase fabric survivability requirement 1 level.

3. Placement of excessive initial cover material thickness may cause bearing failure of soft subgrades.

TABLE 3—*Minimum geotextile properties.*[a]

Properties	Filtration		Erosion		Separation/Reinforcement			
	Unprotected	Protected	Unprotected	Protected	Very High	High	Moderate	Low
Grab strength, kg (lb) (ASTM D 1682)	80 (180)	36 (80)	89 (200)	40 (90)	120 (270)	80 (180)	58 (130)	40 (90)
Puncture strength, kg (lb) (ASTM D 751-79)	36 (80)	11 (25)	36 (80)	18 (40)	49 (110)	33.5 (75)	18 (40)	13 (30)
Burst strength, kg/cm² (lb/in.²) (ASTM D 751-79)	20 (290)	9 (130)	22 (320)	10 (145)	30 (430)	20 (290)	14.5 (210)	10 (145)
Trapezoidal tear, kg (lb) (ASTM D 1117)	22 (50)	11 (25)	22 (50)	13.4 (30)	33.5 (75)	22 (50)	18 (40)	13.4 (30)
Elongation at failure, (ASTM D 1682)	...	...	20%	20%	...	...	...	...
Protection/rating criteria	Fabric is protected when: Used in drainage trenches or below concrete.		If cushioned by layer of sand or zero height drop.		Reference Table 2 for application of minimum fabric rating.			

[a] AASHTO-AGC-ARTBA Joint Committee Interim Specifications.

space exists behind the fabric. Fabric roll ends should be overlapped a minimum of 30 cm (12 in.) or seamed together.

3. The granular filter material placed within the trench should be compacted using a vibratory roller to a minimum compaction of 95% Standard Proctor. Compaction seats the fabric granular system against the natural subgrade to reduce voids and minimize settlement to the shoulder area.

4. After placement of the granular fill, the two edges of the geotextile protruding at the top of the trench are overlapped on top of the granular filter material, and then soil or other materials should be placed in the trench and compacted to the desired grade.

If settlement criteria dictate the need for densities greater than that defined by Standard Proctor, then care should be taken to ensure that the additional compaction effort does not damage the geotextile. In these situations, geotextiles having a higher resistance to puncture should be used (see Tables 2 and 3).

Reverse filters commonly associated with erosion control require construction specifications that reflect the need to control the flow of water over the surface of the fabric. As such, specific construction criteria should include the following:

1. Slopes in excess of 2.5 to 1 should not be used.

2. Slopes should be graded to provide a smooth, fairly level surface. The fabric should be laid with the machine direction of the fabric placed parallel to the slope. Folds and wrinkles of the fabric should be avoided.

3. Adjacent rolls should be overlapped 30 cm (12 in.) or should be seamed. A J-type seam using dual seams is preferred for these applications. Overlapped seams should be secured using metal pins.

The Corps of Engineers specify steel securing pins, nominally 0.5 cm (3/16 in.) in diameter, 46 cm (18 in.) long, pointed at one end and fitted with 3.8-cm (1.5-in.)-diameter washers at the other end for use in securing fabrics in firm soils. The pin spacing is a function of the slope with the following spacings recommended:

Slope	Pin Spacing Per Row
3:1	0.6 m (2 ft)
3:1 to 4:1	0.9 m (3 ft)
4:1	1.5 m (5 ft)
All slopes	1.8 m (6 ft) between parallel rows of pins

4. The placement of stone cover should begin at the base of the slope and at the center of the geotextile-covered zone. The placement of the cover material should result in the tensioning of the underlying geotextile.

5. For fabrics having properties exceeding that required for protected applications (see Table 3) but with no cushion, the height of drop for stones less than 110 kg (250 lb) should be placed without free-fall.

6. For "protected" fabrics with a cushion and fabrics having properties exceeding that for "unprotected" applications, the height of drop for stones less than 110 kg (250 lb) should be less than 0.9 m (3 ft); stones greater than 110 kg (250 lb) should be placed with no free-fall.

7. If stones greater than 110 kg (250 lb) must be dropped, or if a height of drop exceeding 0.9 m (3 ft) must be used, then field trials should be performed to determine the maximum height of safe drop without damaging the fabric.

8. Stones weighing in excess of 45 kg (100 lb) should not be allowed to roll along the surface of the fabric.
9. Contouring of the stones should be achieved during their initial placement, with no grading of the stones after placement allowed.

Erosion control fabrics can be easily damaged both by exceeding height-of-drop criteria and by placing or moving the cover stones in such a manner that the pinned joints of the fabric open. For these reasons it is important that a qualified geotechnical engineer monitor the installation of such fabrics.

Silt fences constructed with geotextiles were extensively studied by the Virginia Department Transportation after an initial study indicated that less than 20% of such installations were effective. Based on these studies [*4–6*], the following installation procedure is currently recommended for silt fences constructed using geotextiles:

1. The height of the filter fabric silt fence should not exceed 0.9 m (3 ft).
2. The filter fabric should be purchased in a continuous roll and cut to the length of the barrier to avoid the use of joints.
3. Wood or steel posts are set securely on line no more than 3 m (10 ft) apart. Wood posts should be at least 7.6 cm (3 in.) in diameter and a minimum of 1.5 m (5 ft) long. Metal posts should be T-shaped, weighing more than 2.0 kg/m (1.33 lb/ft), measuring no less than 1.5 m (5 ft) long, and having projections for fastening the wire to the fence.
4. A trench is excavated 10 cm (4 in.) wide and 10 cm (4 in.) deep along the line of posts and upslope from the barrier.
5. When a wire mesh support fence is used, the wire should be a minimum of 14.5-gage woven wire and must be securely fastened to the upslope side of the posts. The wire should extend flush to the ground.
6. The filter fabric is stapled or wired to the fence with the fabric being allowed to extend to the bottom of the trench.
7. With wooden posts, the wire staples must be No. 9 and at least 3.8 cm (1.5 in.) long. Where steel posts are used, 17-gage wire is used in lieu of wire staples.
8. The trench is backfilled and the soil compacted over the filter fabric.
9. If a filter fabric silt fence is to be constructed across a ditch line or drainageway, the barrier should be of sufficient length to eliminate end flow and the plan configuration should resemble a horseshoe, with the ends pointing upslope.

The general sequence in the construction of geotextile silt fences is shown on Fig. 1.

Construction Criteria—Separation

The separation role performed by a geotextile is to prevent adjacent soil layers from moving together. This follows a basic tenet in geotechnical engineering [*7*] that soils with vastly different particle sizes and particle size distributions cannot be placed together. Similarly, if the strength of the finer soil is low in relation to the stresses, the larger-size distributions cannot be placed together. Similarly, if the strength of the finer soil is low in relation to the stresses the larger-size material can exert against it, the larger-size soil will penetrate into the finer material. Geotextiles act to distribute the more concentrated pressures applied by the larger-size material to minimize local bearing failures and at the same time prevent migration of the finer particles through the coarser matrix. It is typical for geotextiles providing separation that some additional reinforcement role may also be provided.

Geotextiles as separators are being used in applications including paved and unpaved roadways, in developing temporary working platforms over weak subgrades, in both new and exist-

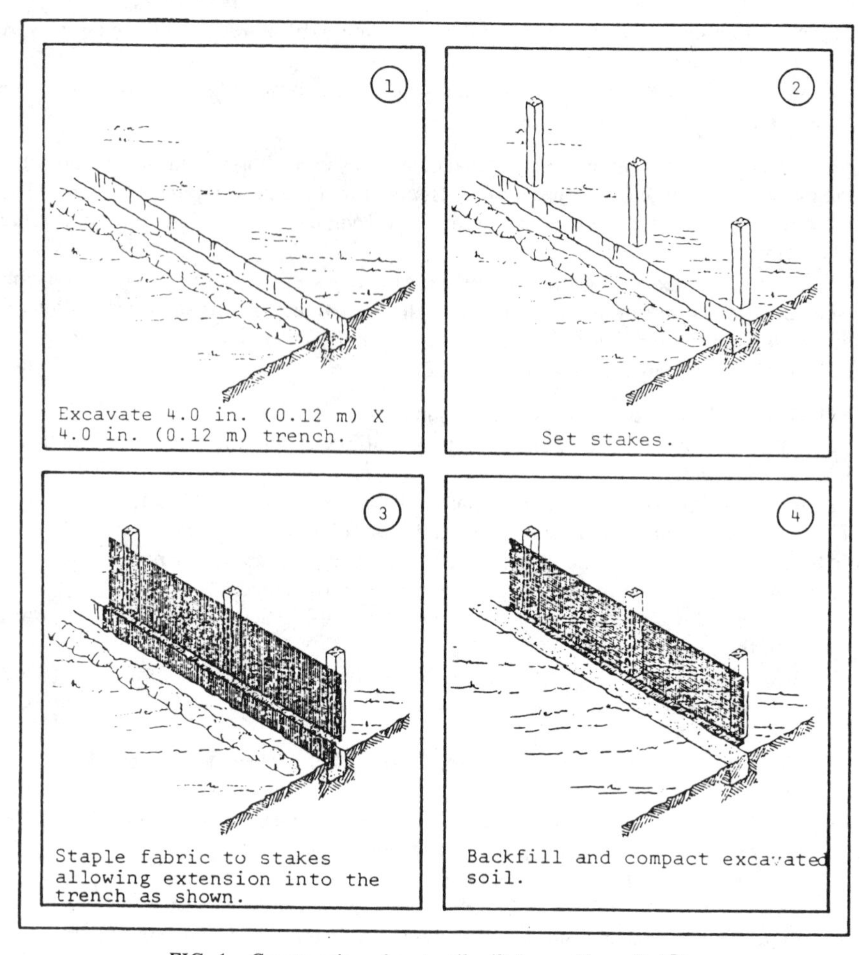

FIG. 1—*Construction of geotextile silt fences (from Ref* 5*).*

ing railroad track installations to prevent ballast fouling, and in new areas such as separating landfill covers from the underlying fill material.

Typical construction problems related to the placement of a separator layer include the following:

1. Ripping or puncturing the fabric during the construction activity.
2. Insufficient cover over the fabric during placement of the cover material, resulting in failure of the subgrade due to excessive pressure generated by construction equipment.
3. Rutting of the underlying subgrade prior to placement of the geotextile, resulting in excessive deformation of the fabric during placement of the cover material.
4. Placing lift thicknesses or dumping the cover material in piles that produce contact stresses at the subgrade that exceed the bearing capacity of the soil.

The general construction procedure for a separator system is shown in Fig. 2. If the geotextile is to provide a reinforcing role in addition to providing separation, then construction criteria for that role should be reviewed and incorporated into the project specifications.

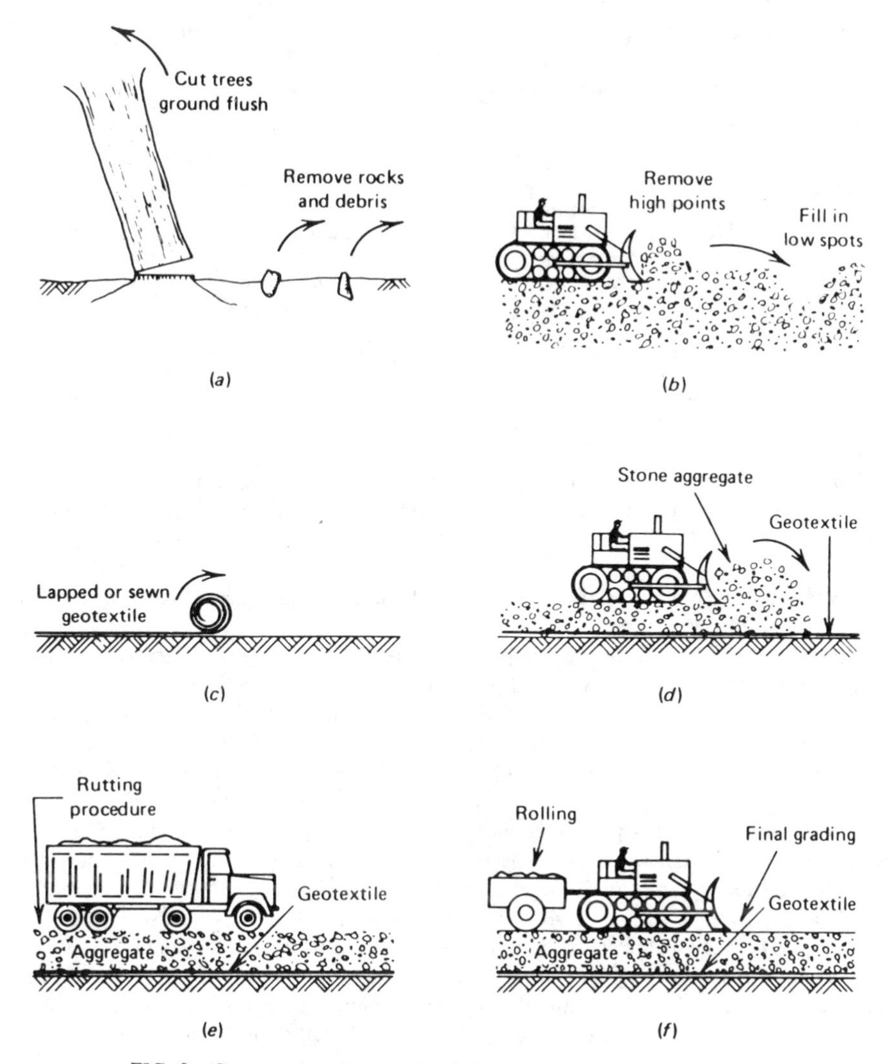

FIG. 2—*Construction of separation/reinforcement system (from Ref 7).*

Construction criteria recommended for a geotextile separator system in typical roadways are as follows [*2*]:

1. The site should be cleared, grubbed, and overexcavated to design grade, taking care to strip all topsoil, soft soils, or other unsuitable material. Isolated pockets where overexcavation is required should be pitched or backfilled to promote positive drainage.
2. During stripping operations, care should be taken not to disturb the subgrade. This may require use of lightweight dozers, etc. for low-strength soils such as saturated cohesionless or low-cohesion soils. Subgrade preparation should reflect the survivability criteria presented in Table 2.
3. Once the subgrade along a particular segment of road alignment has been prepared, the geotextile is unrolled in line with the placement of the new aggregate. The fabric should not be

dragged across the subgrade, and the entire fabric should be placed and rolled out as smoothly as possible.

4. Parallel rolls of fabric should be overlapped or sewn as required. Recommended minimum overlaps are as follows:

California Bearing Ratio, ASTM D 4429[a]	Minimum Overlap
Greater than 2	0.3 to 0.45 m (1 to 1.5 ft)
1 to 2	0.6 to 0.9 m (2 to 3 ft)
0.5 to 1	0.9 m (3 ft)
Less than 0.5	Sewn only
All roll ends	0.9 m (3 ft) or sewn

[a] ASTM Test Method for Bearing Ratio of Soils in Place (D 4429-84).

The fabric should be pinned at all overlaps to maintain them during construction activities.

5. Fabric widths should be selected such that overlaps of parallel rolls occur at the centerline and at the shoulder. Overlaps should not be placed along anticipated main wheel path locations.

6. Overlaps at the end of rolls should be in the direction of the aggregate placement with the previous roll on top.

7. When fabric intersects an existing pavement area, the fabric should extend to the edge of the old system and consideration should be given to anchoring the end of the fabric.

8. The subbase aggregate should be end-dumped on the fabric from the edge of the fabric over previously placed aggregate to eliminate equipment contact with the fabric. For very weak subgrades, the pile heights should be limited to prevent construction-induced road failure.

9. Before covering, the condition of the fabric should be observed by a qualified engineer to determine that no holes or rips exist in the fabric. All such occurrences should be repaired by placing a new layer of fabric extending beyond the defect in all directions at a distance equal to the minimum overlap required for adjacent rolls.

10. The first layer of aggregate should be graded down from the previously placed fill to a thickness of 30 cm (12 in.) or the maximum design thickness. At no time should equipment be allowed on the fabric with less than 20 cm (8 in.) of aggregate between the wheels and the fabric. Construction vehicles should be limited in size and weight such that rutting in the initial lift is less than 7.6 cm (3 in.). If rut depths exceed 7.6 cm (3 in.), then the size and weight of the equipment must be reduced.

11. The first lift should be compacted by "trucking" with a dozer and then compacted with a smooth-drum vibratory roller to obtain a minimum compacted density. For soft soils the design density of the first layer should be 5% less than for the remaining layers. Use of the vibratory roller should be closely monitored in applications where the existing subgrade consists of saturated, loose cohesionless soils. Liquefaction of these soils due to excess pore water pressures can lead to bearing capacity failure of the subgrade.

12. Turning of construction equipment should not be permitted on the first layer of aggregate placed upon the fabric.

13. Any ruts that form during construction should be filled with new aggregate to maintain an adequate depth of cover over the fabric as shown in Fig. 3. In no case should ruts be bladed down as this decreases the amount of aggregate cover between the ruts.

14. All remaining lifts of subbase aggregate should be placed in lifts not exceeding 23 cm (9-in.) loose thickness.

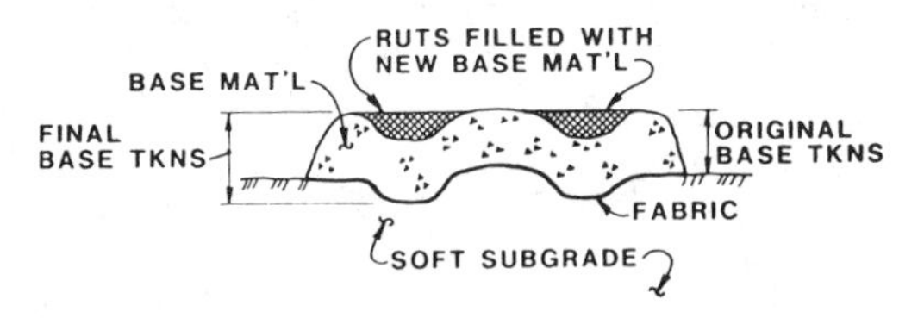

FIG. 3—*Repair of rutting with additional base material (from Ref* 8*).*

If the geotextile is to provide additional roles beyond separation, then construction criteria for those roles should be reviewed and incorporated into the project specifications.

Construction Criteria—Reinforcement

Geotextile systems designed to use the fabric in a reinforcement mode include geotextile-reinforced retaining walls, support of embankments built upon soft soils, and soil encapsulation. The basic premise to all design is for the fabric to either supply tensile capacity lacking in the unreinforced soil or to apply a pseudoconfining pressure that results in an increase in the ultimate strength of the soil.

Construction problems in the placement of a reinforcement layer typically relate to conditions that reduce the tensile capacity of the fabric. Some of these construction problems include:

1. Ripping or puncturing the fabric during construction leading to failure of the fabric.
2. Failure of joining seams.
3. Failure of the fabric itself in tension.

An additional failure mode associated with embankments built upon soft subgrades is localized bearing capacity failure of the subgrade.

A clear definition of construction criteria for embankments built upon soft soils is important since the desired fill placement procedure is not the same as for conventional embankments. For this category of structure, the construction criteria include the following:

1. The site should be prepared to the most exhaustive criteria defined in Table 2 per fabric survivability. Large depressions should be lined with the geotextile with 1.8 m (6 ft) of material projecting beyond the depression. The depression should be filled with granular material level to the surrounding surface prior to placement of the primary geotextile layer.

2. Seams should be of the "butterfly" or "J" type as shown in Fig. 4. A lock-type stitch should be used with double seams no farther than 6 to 12 mm (1/4 to 1/2 in.) apart as required for all critical applications. A successful seam will provide no more than two-thirds of the fabric tensile strength. All seams should be tested to ensure that they meet the design strength criteria.

3. Before covering, the fabric should be inspected by a qualified engineer for damage and to ensure that the fabric is rolled out as smoothly as possible. All wrinkles and folds in the fabric should be removed by stretching and staking as required.

4. Placement of the first layer of fill upon the geotextile must follow the guidelines previously presented herein for the separation mode. It is particularly important that the rut depth be maintained at less than 7.6 cm (3 in.) and that fresh stone material should be added to all ruts filled between passes.

5. Heavy compaction equipment should not be used on the first layer of fill and vibratory rollers should be avoided if the underlying subgrade is a saturated, loose cohesionless soil.

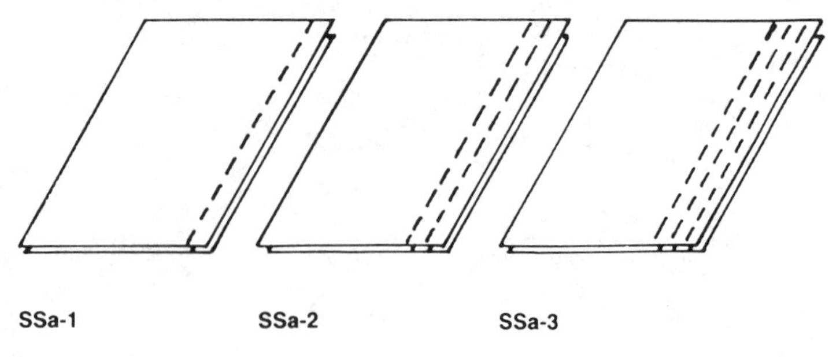

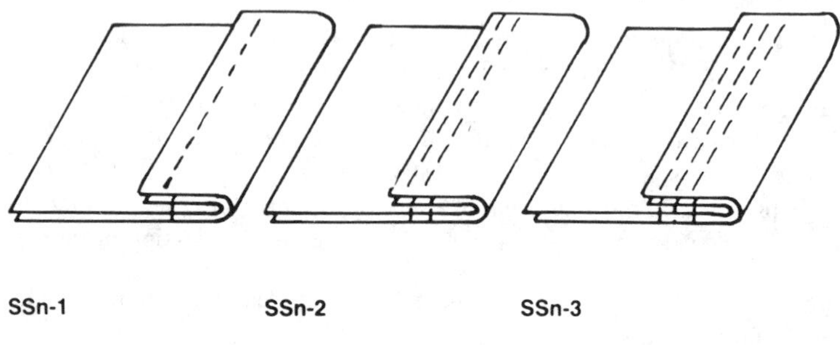

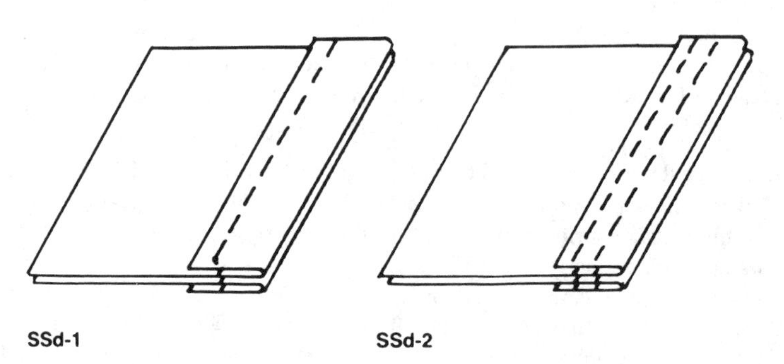

FIG. 4—*Geotextile seaming techniques (from Ref* 9*).*

6. Fill placement for embankments placed upon soft soils must follow the scheme shown in Fig. 5. This scheme places fill first on the edges of the fabric to anchor the fabric and then tensions or "sets" the fabric by placing the center fill.

Embankments built upon subgrades so soft that mud waves develop require controls beyond that described here [*10–12*].

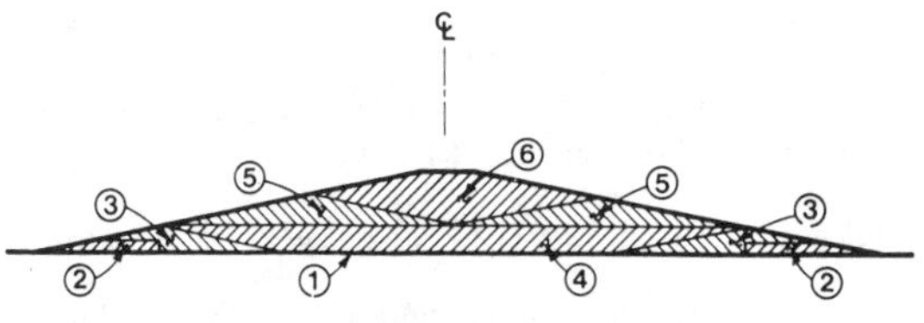

SEQUENCE OF CONSTRUCTION

1. LAY FABRIC IN CONTINUOUS TRANSVERSE STRIPS, SEW STRIPS TOGETHER.
2. END DUMP ACCESS ROADS.
3. CONSTRUCT OUTSIDE SECTIONS TO ANCHOR FABRIC
4. CONSTRUCT INTERIOR SECTION TO "SET" FABRIC.
5. CONSTRUCT INTERMEDIATE SECTIONS TO TENSION FABRIC.
6. CONSTRUCT FINAL CENTER SECTION.

FIG. 5—*Recommended construction sequence for fabric-reinforced embankments (from Ref* 1*).*

Retaining walls constructed using fabric reinforcement must build each layer using criteria similar to that developed for the first fill layer in separation or reinforcement applications. In addition, it may be desirable to specify that all lifts be placed at a 1% slope contoured to provide drainage away from the face of the wall. This minimizes the potential for ponding due to heavy rains during construction.

Summary

Construction criteria have been presented for use in developing each of the four primary roles of geotextiles. An important criteria common to all roles is that fabrics susceptible to degradation from ultraviolet light (UV) should be placed and covered within five days and those fabrics that are UV stabilized should be placed and covered within 30 days. Applications such as silt fences must be constructed with the understanding that all fabrics degrade in direct sunlight so that such structures are temporary measures at best.

An additional common denominator to all function modes is the need to have a qualified engineer present in the field to ensure that the fabric and the first layer of fill are properly placed. A major difficulty here is defining what a "qualified" engineer should be since many geotechnical engineers are not familiar with geotextiles and their design or installation.

References

[1] Bell, J. R., Hicks, R. G., et al., "Evaluation of Test Methods and Use Criteria for Geotechnical Fabrics in Highway Application," Report No. FHWA/RD-80-021, Oregon State University, Corvallis, 1980.

[2] Holtz, R. D. and Christopher, B., "Geotextile Engineering Manual," prepared for Federal Highway Administration, Washington, DC, 1984.

[3] AASHTO-AGC-ARTBA Joint Committee Interim Specifications.

[4] Wyant, D. C., "Evaluation of Filter Fabrics For Use As Silt Fences," Virginia Highway and Transportation Research Council, Charlottesville, VA, 1980.

[5] "Erosion and Sediment Control Manual," Virginia Department of Highways and Transportation, Charlottesville, VA, 1984.

[6] Sherwood, W. C. and Wyant, D. C., "Installation of Straw Barriers and Silt Fences," Report No. VHIRC77-R18, Virginia Highway and Transportation Research Council, Charlottesville, VA, 1976.

[7] Koerner, R. M., *Construction and Geotechnical Methods in Foundation Engineering*, McGraw-Hill Book Co., New York, 1984.
[8] Haliburton, T. A., Lawmaster, J. D., and McGuffey, V. E., "Use of Engineering Fabrics in Transportation Related Applications," Final Report Under Contrast No. DRFH-80-C-0094, Federal Highway Administration, Washington, DC, 1982.
[9] Diaz, V., "Thread Selection for Geotextiles," *Geotechnical Fabrics Report*, Vol. 3, No. 1, Jan.-Feb. 1985.
[10] Haliburton, T. A., "Design of Test Section for Pinto Pass Dike, Mobile, Alabama," report to U.S. Army Engineer District, Mobile, Alabama, by Haliburton and Associates, Stillwater, Oklahoma, under Contract No. DACW01-78-C-0092.
[11] Haliburton, T. A., Anglin, C. C., and Lawmaster, J. D., "Selection of Geotechnical Fabrics for Embankment Reinforcement," report to U.S. Army Engineer District, Mobile, Alabama, Oklahoma State University, Stillwater, Oklahoma, 1978.
[12] Haliburton, T. A., Fowler, J., and Langna, J. P., "Design and Construction of Fabric Reinforced Embankment Test Section at Pinto Pass, Mobile, Alabama," Transportation Research Record No. 749, Transportation Research Board, National Academy of Sciences, Washington, DC, 1980.

The Future of Geotextiles in Geotechnical Engineering

Barry R. Christopher[1]

Tomorrow's Tests for Geotextile Characterization and Evaluation

REFERENCE: Christopher, B. R., **"Tomorrow's Tests for Geotextile Characterization and Evaluation,"** *Geotextile Testing and the Design Engineer, ASTM STP 952*, J. E. Fluet, Jr., Ed., American Society for Testing and Materials, Philadelphia, 1987, pp. 141–144.

ABSTRACT: Tomorrow's tests for geotextiles will be developed around today's design needs and tomorrow's design requirements. In this paper, current testing needs are reviewed and future test requirements are postulated. Current test needs that are reviewed include: standardized tests for basic fabric properties; standardized performance tests for soil-fabric properties; model studies to evaluate the mechanisms which enhance fabric performance; and studies to define the relationship between physical-mechanical properties and soil-fabric behavior. These needs are discussed in relation to providing the framework for future test development. Future advancements in testing technology are reviewed with respect to modification of existing test methods to provide better characterization of geotextiles and to provide answers to current design questions. Potential test requirements for new applications, new design methods, and new products are postulated. Finally, test equipment development that may be required to meet the needs of advancing test technology is reviewed.

KEY WORDS: geotextiles, properties, soil-fabric properties, laboratory tests, standardization, instrumentation, research

Future tests will be required to provide input parameters for tomorrow's designs. However, before proceeding to the future, a concentrated effort must be placed on current test requirements to meet today's design needs. Current testing concerns include:

1. Standardized index tests for basic fabric properties.
2. Standardized performance tests for soil-fabric interaction evaluation.
3. Model studies to evaluate the mechanisms which enhance fabric performance.
4. Development of definitive relationships between physical-mechanical geotextile properties and soil-fabric behavior.

Meeting these testing requirements will lay a firm foundation for the development of tomorrow's tests.

Tomorrow's tests will center around:

1. Modification of existing tests to provide realistic assessments of geotextile characteristics.
2. Tests required to answer design questions.
3. New tests for new applications.
4. New tests for new materials.

Current testing needs and future advancements required to meet these needs will be discussed in the following two sections.

[1] Principal engineer, STS Consultants, Ltd., Northbrook, IL 60062.

Current Test Needs

One of the most substantial needs in testing geotextiles is the standardization of existing test procedures to define basic geotextile properties. Physical properties are the means of communication between the design engineer and producers as well as other designers. Currently numerous test procedures are being used by various agencies as well as by private engineering groups to evaluate the same geotextile property. Procedural variations between test methods typically result in significant differences in the property values obtained. Such discrepancies make it difficult, if not impossible, to compare project performance or to ascertain the validity of design criteria used by each organization. Thus, a vast amount of knowledge is being lost that could be shared if a common set of test standards were available. In addition, nonstandardization leads to poor communication between the design engineer and geotextile manufacturers and suppliers as to exactly what the real geotextile requirements are for specific applications. Poor communication thus hampers the development of new materials to meet engineering needs. The efforts of ASTM Committee D35 on Geotextiles, Geomembranes, and Related Products are aimed directly at solving this problem. However, cooperation between all organizations and more effective communication between users are essential to rectify the situation in the near future.

Another similar existing need is the standardization of soil-fabric performance tests. In-soil tests provide the best means of quantitatively assessing anticipated field performance. Several soil-fabric test methods have been proposed over the past decade for determination of in-soil geotextile strength, soil-fabric friction, in-soil creep response, and filtration potential. Adoption and standardization of these methods would vastly improve performance evaluation. However, the geotechnical community must understand that it is their responsibility to both develop and perform these tests. Performance tests can only be performed by experienced qualified geotechnical laboratories under the user's directions. Generally, such tests must be performed using site-specific soils. These capabilities and facilities are generally far beyond those available to most geotextile manufacturers and suppliers. The user must thus insist that his laboratory gear up to perform these tests, the same as they currently do for other geotechnical design-type tests.

Once test standards have been established, laboratory models and field trials need to be performed to develop definitive relationships between actual fabric performance and physical-mechanical geotextile property values. Construction equipment needs to be run over various types of aggregate placed over geotextiles, and rocks need to be dropped on fabrics from various heights to relate such properties as grab strength, tear strength, burst strength, puncture strength, and impact resistance to survivability requirements.

Research is also needed to establish design methodology for many of today's applications. Systems' response of detailed laboratory and field models needs to be analyzed to verify existing analytical design methods and to develop new design approaches. For example, reinforced soil embankment models could be used to analyze such items as stress redistribution resulting from soil-fabric interaction and assumptions concerning stress relaxation versus geotextile creep. Laboratory and field models may also be used to develop both qualitative and quantitative evaluation methods for new applications. The engineering community needs a dedicated effort from universities and research organizations to seek funding and to perform such studies to advance the state of the practice in using geotextiles by providing these needed design tools.

A concentrated effort by the geotechnical and geotextile community in meeting today's design and testing needs will lead to the development of tomorrow's tests.

Tomorrow's Tests

In line with the current needs, the first new tests should be developed to provide answers to existing design questions. Tests to provide input parameters for analytical evaluation methods need to be developed where currently no reliable test methods exist. This would also include modification and improvement of existing tests to provide a more realistic assessment of geotextile characteristics as they relate to specific design requirements. Table 1 lists many of the properties where reliable means of evaluation do not currently exist.

The basic and fundamental research previously mentioned, including the use of laboratory and field models, may provide the tools for developing reliable test methods for evaluating the properties listed in Table 1.

Other tests required to meet current design needs include field quality control and installation assessment tests. At this time, specimens must be sent to the laboratory several days in advance of actual installation. In many cases, data may not be available until after installation. Tests need to be developed that can be easily performed in the field as geotextiles are delivered to provide a rapid determination of material properties prior to installation. Such tests would be especially important for critical projects. As an example, field tests might include spring-loaded impact or puncture devices.

In addition, field tests should be developed that will allow for an in-situ assessment of the installation. Both destructive and nondestructive procedures should be examined. Such methods may include electronic and mechanical instrumentation to evaluate stress-strain and creep response. Sonic wave or dynamic evaluation techniques may provide indications of pretension moduli. For systems performance, deflectometers may provide a means of measuring moduli improvement provided by geosynthetic reinforced aggregate systems for both design and long-term assessment. Nondestructive tests to evaluate seam integrity and effectiveness would also be extremely valuable.

As new applications and design methods are developed, new tests will be required for design evaluation. Future design methods will include the use of computers and sophisticated numerical techniques such as finite-element and finite-difference analysis. Such design methods are only as effective as the input parameters, which in many cases are as of yet undefined. For example, in-soil stress-path tests will have to be developed to provide input for reinforcement analytical models.

New applications will, of course, introduce new test requirements. For example, future reinforced roadway applications will most assuredly include pretensioning concepts which will require evaluation of such undefined parameters as dynamic strength, dynamic creep response, and stress relaxation. Future designs will also include new construction techniques that may require special test development. For the aforementioned prestressed reinforced roadway design, special construction techniques will be required to provide pretensioning, and test methods will be required to evaluate its effectiveness. Other applications may include reinforcement of structural fill to increase foundation support for buildings, replacement of core drains for

TABLE 1—*Properties required to meet today's design requirements.*

A. Seam strength	L. Fabric roughness in relation to friction angle
B. Creep resistance	M. Soil-Fabric shear strength
C. Poisson's ratio	N. In-Soil geotextile anchorage requirements
D. Confined stress-strain characteristics	O. Chemical and biological resistance
E. Dynamic modulus and strength	P. Poremetry (pore size distribution)
F. Dynamic creep characteristics	Q. Permeability under load
G. Flexibility	R. Intrinsic gradient required to initiate flow
H. Abrasion resistance	S. Permeability and flow capacity under partially saturated conditions
I. Penetration resistance	T. Reversing flow effects on retention and filtration
J. Fabric cutting resistance	
K. Impact resistance	

earth dams, and, of course, providing subgrade stabilization for construction of lunar base stations.

Along with new applications will come the development of new geosynthetics. As communication between designers and suppliers improve, manufacturers will be able to modify and develop products to better meet design needs. A good example of new materials are the improved prefabricated drainage products used to replace drainage aggregate. New products will also be developed with advances in polymeric and textile technology. Highly technical materials will become less costly to produce and will thus provide cost-effective, alternative, high-strength geosynthetics. New tests, or at least modification of existing tests, will most likely be required so that procedures are effective in providing appropriate property values for these new materials.

Conclusions

New design methods, applications, and geotextile materials are continually being developed that provide valuable design alternatives for civil engineers. Test methodology provides the tools for assessing both the material characteristics and their effectiveness for a particular application. Engineers who are ultimately responsible for the design cannot design with confidence unless the needed tools are provided. Today, test technology lags seriously behind the use of geosynthetics. This is largely a problem with users' acceptance and use of test methods. "Using" means actually running the test in their own laboratories to compare properties with design performance. The results of committees such as ASTM D-35 in meeting current test requirements and future test needs will not be realized until the future, but the user must start using available tools today.

Today's questions concerning this paper will unfortunately have to wait until tomorrow for the answers.

J. P. Giroud[1]

Tomorrow's Designs for Geotextile Applications

REFERENCE: Giroud, J. P., **"Tomorrow's Designs for Geotextile Applications,"** *Geotextile Testing and the Design Engineer, ASTM STP 952,* J. E. Fluet, Jr., Ed., American Society for Testing and Materials, Philadelphia, 1987, pp. 145–158.

ABSTRACT: The first part of this paper presents a review of existing design methods for selected geotextile applications such as filtration, drainage, and unpaved roads. The discussion shows that geotextile applications can be designed using rational methods and that most of these methods are similar to classical soil mechanics design methods. Therefore, geotechnical engineers are well-equipped with the basics for designing with geotextiles, and designs that one would think are "tomorrow's designs" can, in fact, be done as soon as tomorrow morning by those geotechnical engineers who are aware of available design methods. The second part of this paper is an attempt at predicting the evolution of design concepts and design methods. The following topics are addressed: new materials, new construction methods, progress in the evaluation of geotextile properties, and expected results from research. The discussion shows how important it is for geotechnical engineers to keep abreast of new developments.

KEY WORDS: design, drainage, filtration, future developments, geotechnical engineering, geotextiles, geosynthetics, soil, soil reinforcement, unpaved roads

The first idea that came to mind when preparing this paper was to predict how we will design with geotextiles in the future. Such an attempt might be useful and will therefore be discussed later in this paper with a view to opening up new avenues to design engineers and to setting long-term goals for researchers.

But the engineering profession has more immediate needs. Many design engineers are anxious to begin using geotextiles now, if only because they see a growing number of their peers using geotextiles to prepare designs that are more elegant, more economical, and easier to construct. Many design engineers have a geotextile design problem to solve tomorrow morning, not one morning in 1999. The first part of this paper on "tomorrow's designs" will therefore be devoted to "tomorrow morning's designs," that is, a discussion of design methods readily available and usable as early as tomorrow morning.

One of the main goals of the ASTM Symposium on Geotextile Testing and the Design Engineer and this publication is to make engineers more aware of the extent to which many geotextile applications can be designed using rational methods. More importantly, most of these methods are similar to classical soil mechanics methods. Geotechnical engineers are, therefore, equipped with the basics required to design with geotextiles. But, to actually design, the basics are not sufficient, and the geotechnical engineer needs practical information on design methods. This need is addressed in following paragraphs for three typical applications.

[1]Principal engineer, GeoServices, Inc., Boynton Beach, FL 33435.

Designing with Geotextiles Tomorrow Morning

For many geotechnical engineers, manufacturers' brochures are presently the sole source of design information. Some of the brochures provide useful information, and all contain performance claims that are more or less justified. Geotechnical engineers trying to design a geotextile application will be either biased if they use only one brochure or confused by conflicting claims if they use more than one.

A rational design approach including the following two steps is necessary for proper geotextile design:

1. Identify the mechanisms through which the geotextile is expected to perform in the considered application, and identify geotextile properties governing these mechanisms (selection of appropriate test methods to evaluate these properties would be an important part of this step; test methods are not discussed in this paper).
2. Determine the required values of the geotextile properties by using a method of design.

The phrase "design by function" is often used to designate the just-described approach. This is in reference to the classical geotextile functions such as filtration, fluid transmission, separation, and reinforcement. The phrase "design by mechanism" would be more appropriate since each function may involve several mechanisms (for example, the filtration function involves three mechanisms while the separation function involves two, as discussed hereafter).

The two-step approach just discussed will be used in following paragraphs for three typical applications: geotextile filters, geosynthetic drains, and geosynthetic-reinforced unpaved roads. Design methods are also discussed in Ref *1* and in papers in this publication, such as Ref *2* for geosynthetic drains and Ref *3* for reinforcement of steep slopes, earth walls, and embankments.

The following discussion of three typical applications illustrates the parallelism between design methods used for geotextile applications and classical soil mechanics design methods. This parallelism would be similarly illustrated by any other geotextile applications.

Design of Geotextile Filters

Mechanisms and Relevant Properties—When geotextiles are used as filters between soils and drainage materials (gravel, perforated or slotted pipes, synthetic drainage layers) they: (1) must allow water to pass without building up excess pore water pressure; (2) must prevent soil particles from moving (with the exception of a few small particles adjacent to the filter, which may be carried away by the water); and (3) must not become clogged by soil particles.

Identification of geotextile properties which are relevant to either of the first two mechanisms just indicated is straightforward: the first mechanism requires a geotextile with high permeability (therefore with large openings), while the second mechanism requires a geotextile with small openings. A compromise must be found between these two contradictory requirements, and this is addressed in the subsequent discussion on design methods.

The potential for clogging depends on the soil and on the geotextile. With certain soils (such as soils that are gap-graded and/or in a loose state) where fine particles can easily move, clogging is more likely to occur than with other soils, regardless of the type of filter (sand or geotextile). In such cases, it is recommended that a laboratory test simulating actual conditions, such as a "gradient ratio test" [*4*], be conducted. However, with most soils encountered in civil engineering projects (that is, soils with a continuous grading curve and in a dense state), clogging can be avoided by proper selection of the geotextile filter. The geotextile property relevant to the nonclogging condition is not as obvious as the geotextile properties related to the two previous mechanisms. This property can be identified by considering the extreme case of a filter made of a perforated sheet with a small number of perforations per unit area. Such a filter could be satisfactory with respect to the first two mechanisms just presented: it could have enough permeability to allow flow of water without excessive pore pressure, and its perforations could be small enough to retain soil particles. However, flow of water would have to concentrate in the vicinity of the perforations (Fig. 1). As a result, flow velocity would increase and the soil particles would be carried toward the perforations. A few particles would pass, but a large number would accumulate, thereby generating clogging. Therefore, the geotextile property related to the nonclogging condition is that the geotextile should have a sufficient number of openings per unit area so that the clogging mechanism just described will not occur.

In the case of a woven geotextile, this condition can be expressed by specifying a minimum "percent open area" (ratio, expressed as a percentage, between total surface area of openings and total surface area of a woven geotextile) in addition to the maximum opening size already mentioned. The minimum percent open area alone does not indicate that there should be a minimum number of openings, but it does so when it is combined with the requirement regarding the maximum opening size.

In the case of a nonwoven geotextile, the same requirement can be expressed by specifying a minimum porosity (ratio, which can be expressed as a percentage, between the volume of the voids and the total volume of the geotextile). It is possible to show that, in a randomly organized porous structure (which is approximately the case of a nonwoven geotextile), the percent open area of a cross section of the structure is equal to the porosity.

In conclusion, the properties relevant to the three mechanisms just described are: (1) permeability; (2) opening size; and (3) percent open area (for wovens) or porosity (for nonwovens).

Design Methods—Since there are three mechanisms through which a geotextile performs the filtration function, there should be three criteria for the relevant geotextile properties:

1. Criterion for minimum permeability ("permeability criterion")

$$k_g \geq \lambda k_s \quad (1)$$

where k_g, k_s = hydraulic conductivity (coefficient of permeability) of the geotextile and the soil, respectively. A theoretical analysis of pore pressures behind filters [*5*] gives an analytical expression for λ as a function of the thickness of the filter, with a value $\lambda = 0.1$ for typical geotextiles. (Note: the same analytical expression would give $\lambda = 10$ for typical granular filters.) Since the hydraulic conductivity of geotextiles typically decreases by at least one order of magnitude under compressive stresses, it is recommended practically to take $\lambda = 1$ or 10 [*6*].

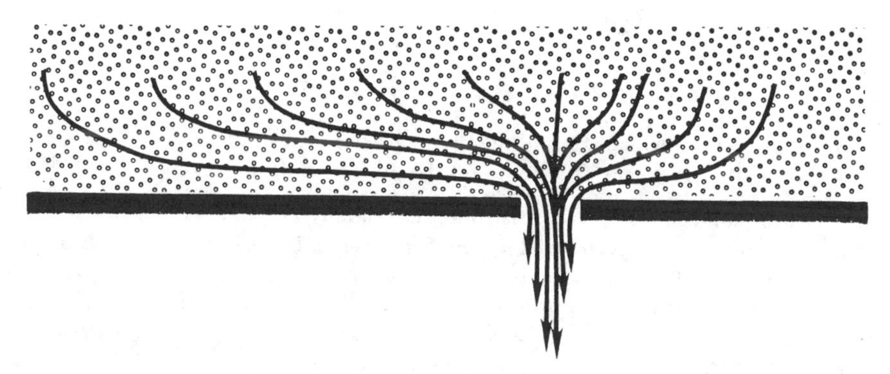

FIG. 1—*Flow concentration in the case of a filter with a small number of openings.*

2. Criterion for maximum opening size ("retention criterion")

$$O_{95} \leq \mu d_{85} \quad \text{or} \quad \mu' d_{50} \tag{2}$$

where

O_{95} = size which is larger than 95% of the geotextile's openings [Note: an approximate value of O_{95} is given by the apparent opening size (AOS) obtained by sieving calibrated glass beads through the geotextile],
d_{85} (d_{50}) = size which is larger than 85% (50%) of the soil's particle size, and
μ and μ' = dimensionless coefficients.

The value $\mu = 1$ is sometimes used, but this is an oversimplification since μ depends on many parameters such as the soil density, the soil particle size distribution characterized by its coefficient of uniformity, and the flow characterized by the hydraulic gradient. Values for μ have been established using theoretical consideration of the mobility of soil particles [*5*], and other values can be found in Refs *6* and *7*.

3. Criterion for minimum percent open area (wovens) or minimum porosity (nonwovens) ("nonclogging criterion")

$$A \geq A_{min} \quad \text{(wovens)} \tag{3}$$

$$n \geq n_{min} \quad \text{(nonwovens)} \tag{4}$$

where

A = percent open area of a woven geotextile, and
n = porosity of a nonwoven geotextile.

Based on experience, typical values recommended for A_{min} are of the order of 5%. Since nonwoven geotextiles and granular materials have similarities, we have recommended that 30% be used for n_{min} because 30% is a typical porosity for granular filters. This recommendation has been followed by Christopher and Holtz [*6*]. The value of n_{min} should be determined under compressive stresses similar to those encountered in the field.

It is interesting to compare the just-cited geotextile filter criteria to the classical granular filter criteria now given

$$d_{15f} \geq 4 d_{15s} \tag{5}$$

$$d_{15f} \leq 4 d_{85s} \tag{6}$$

where

d_{15f} = size larger than 15% of the filter particles, and
d_{15s}, (d_{85s}) = size larger than 15% (85%) of the soil particles.

Since the hydraulic conductivity (coefficient of permeability) of a granular material is approximately proportional to the square of its d_{15}, Eq 5 means that the coefficient of permeability of the filter should be of the order of ten times the coefficient of permeability of the soil. This is consistent with the theoretical derivation of Eq 1 (see the comment after Eq 1).

Since the opening size (that is, the spacing between particles) of a granular material is approximately one fifth of its d_{15}, Eq 6 means that the opening size of the filter should be approximately smaller than the d_{85} of the soil. Therefore Eq 6 is equivalent to Eq 2.

In conclusion, the permeability and retention criteria for geotextile filters are similar to classical criteria for granular filters. The third criterion used for geotextiles, the "nonclogging criterion" (Eq 4), is also important for granular filters. This criterion is never stated for granular filters, however, because granular filters always have a porosity of the order of 30% or larger, thereby automatically satisfying the third criterion.

Design of Geosynthetic Drains

Mechanisms and Relevant Properties—For years, thick needle-punched nonwoven geotextiles have been used to convey liquids or gases within their plane. In the past few years, new types of geosynthetics with a high drainage capacity have been introduced, such as mats, nets, and waffled plates. These products are called "synthetic drainage layers" and are often available associated with an attached geotextile filter, thus forming a "geocomposite." Such a geocomposite provides the dual functions of drainage (synthetic drainage layer) and filtration (geotextile filter).

The mechanism involved in this application is elementary: the geosynthetic provides an open space through which the liquid or the gas can flow.

Flow of liquid or gas in a needle-punched nonwoven geotextile is usually laminar, like flow in sands. Therefore, the classical Darcy's equation can be used. The two relevant properties of the geotextile are its thickness and its coefficient of permeability (hydraulic conductivity) in the direction of its plane ("in-plane coefficient of permeability" or "in-plane conductivity"). As shown by Darcy's equation, these two parameters can be combined into a single parameter, "transmissivity" (conductivity multiplied by thickness).

Flow of liquid or gas in a synthetic drainage layer is usually turbulent like flow in gravel and, if Darcy's equation is used, the coefficient of permeability and, consequently, the transmissivity, decrease when the gradient increases.

Flow in a synthetic drainage layer is also influenced by compressibility of the synthetic drainage layer and by boundary conditions, as now discussed:

1. Granular drainage materials such as sand and gravel exhibit little compressibility, while all synthetic drainage layers are more or less compressible. Under a compressive stress, the thickness and consequently the transmissivity of a synthetic drainage layer decrease.
2. Since synthetic drainage layers are relatively thin, boundary conditions have a significant influence on flow velocity. Therefore, selection of the geotextile filter is critical not only for filtration characteristics but also regarding the ability of the geotextile to provide a smooth boundary to the flow. Reductions of transmissivity by more than one order of magnitude have been measured for a synthetic drainage layer between two geotextiles as compared to the ideal case of the same drainage layer between two stiff smooth surfaces [5].

In conclusion, the key properties of the synthetic drainage layer itself are transmissivity and compressibility, and geotextile filter selection has a marked influence on the transmissivity. In addition, when a drainage layer is used in a leakage detection system in a waste disposal facility, rapid detection is more important than quantity of flow conveyed and the in-plane conductivity is a relevant property to consider since it governs flow velocity. A detailed discussion of these properties and their evaluation can be found in Ref *8*.

Design Methods—Since Darcy's equation can be used as just indicated, no special design method is required for geosynthetics used as drains. The only special features to take into account when using traditional methods for geosynthetic drains are the factors that affect the transmissivity: (1) gradient; (2) compressive stresses (because of the compressibility of the geosynthetic); and (3) the flow boundary conditions, that is, type of geotextile or geomembrane in contact with the synthetic drainage layer and stiffness of the soil (resulting from soil type, water content, compaction, etc.) that governs the extent to which the soil will push the geotextile or geomembrane into the synthetic drainage layer's openings.

To design with a synthetic drainage layer, it is necessary to have a series of graphs giving the synthetic drainage layer's transmissivity as a function of the compressive stress, the gradient, and the boundary conditions. An example of such a graph is presented in Fig. 2. It is also necessary to verify that the geotextile filter associated with the synthetic drainage layer meets the filter criteria for the soil in contact.

To design with a needle-punched nonwoven geotextile necessitates only the curve of the transmissivity versus compressive stress (gradient has no influence since flow is then laminar). This curve should be established by a laboratory test simulating the actual boundary conditions, especially if the soil in contact with the geotextile has small particles that can penetrate into the geotextile, thereby reducing its transmissivity.

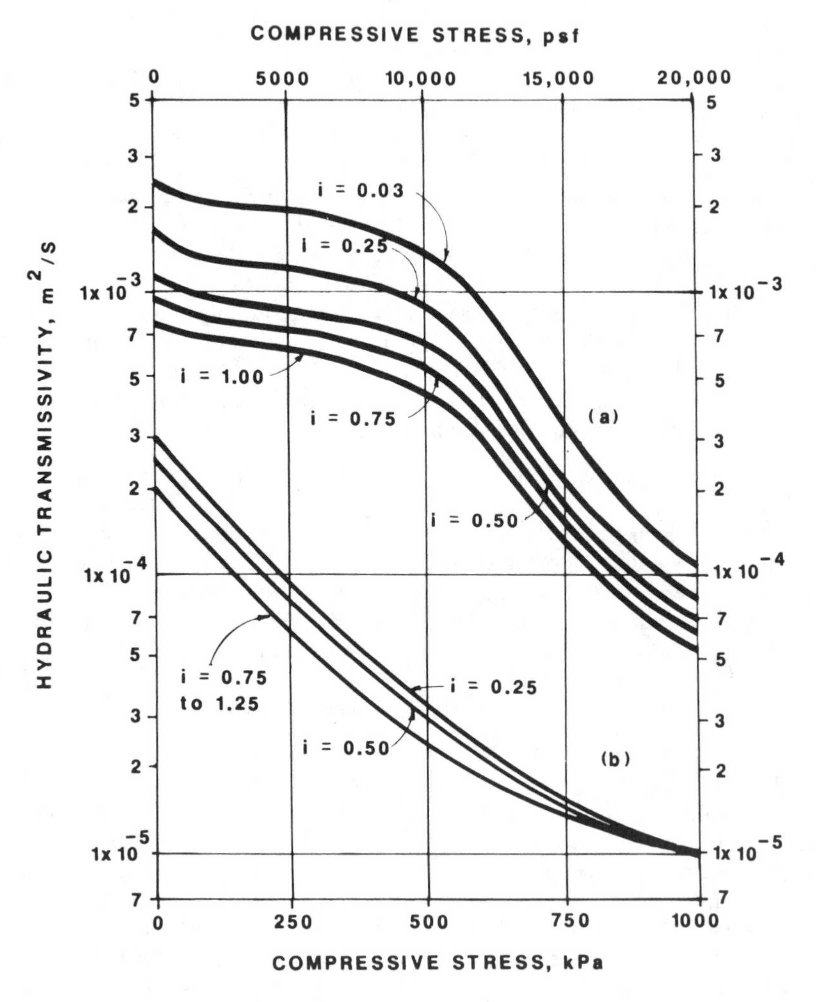

FIG. 2—*Transmissivity of a drainage net as a function of hydraulic gradient and compressive stresses:* (a) *the net is between two geomembranes;* (b) *the net is between a geomembrane and a heat-bonded nonwoven geotextile (Note:* i = *hydraulic gradient). (Data from Ref* 8.*)*

Design methods based on Darcy's equation adapted to geosynthetic drains are presented in Refs *2* and *5*.

Design of Geosynthetic-Reinforced Unpaved Roads

Mechanisms and Relevant Properties—The use of geotextiles in unpaved roads is one of the most widely accepted geotextile applications, and geogrids are increasingly used in this application. Geotextiles and geogrids placed between a soft subgrade soil and an aggregate base layer improve the performance of the unpaved road essentially by performing two functions: separation and reinforcement. Each function involves several mechanisms as discussed in paragraphs that follow.

The geosynthetic separates the subgrade from the base layer through the following two mechanisms:

1. The geosynthetic prevents aggregate from penetrating the subgrade under the action of applied loads.
2. The geosynthetic prevents intrusion of the subgrade soil (usually clay) into the aggregate base layer.

The properties of the geosynthetic which are relevant to the first mechanism are as follows:

1. The opening size should be small enough to retain aggregate. This condition is easily fulfilled by geotextiles and by open structures such as geogrids.
2. The geosynthetic should not be damaged by the aggregate. This can be achieved either by a high resistance to concentrated stresses (evaluated by puncture test or grab test) or by a great extensibility which allows the geosynthetic to follow the shape of the aggregate.

The properties of the geosynthetic which are relevant to the second mechanism are as follows:

1. The opening size should be small enough to prevent fine subgrade particles from intruding into the aggregate. Only geotextiles with rather small openings can fulfill this condition.
2. The geosynthetic should be permeable enough to prevent pore pressure buildup in the often saturated subgrade.

The second function performed by the geosynthetic (geotextile or geogrid) is reinforcement of the unpaved road, which includes three mechanisms:

1. The geosynthetic minimizes lateral spreading of aggregate under the effect of repeated loads. The two relevant properties are the tensile modulus and the interface shear strength (that is, friction and/or interlocking between geosynthetic and aggregate). In fact, a geotextile with a low interface shear strength can have a detrimental effect and facilitate lateral movement of aggregate. This is not a consideration with most geogrids.
2. The geosynthetic provides confinement to the subgrade soil, which restricts subgrade heave between wheels. The relevant property is the tensile modulus.
3. The geosynthetic provides load support under the wheels through the tensioned membrane effect, as indicated in Fig. 3. The relevant property is the tensile modulus. Also, interface shear strength (that is, friction and/or interlocking with aggregate and subgrade soil) is required to provide the necessary lateral anchorage (Fig. 3).

Design Methods—Only the design related to the reinforcement function is discussed here. The three reinforcement mechanisms just presented should be considered.

At the present time, there is no widely available method of design that evaluates the effect of geotextile or geogrid reinforcement on minimizing aggregate lateral spreading.

It has been shown that the confinement effect restricts heave of the subgrade soil and, conse-

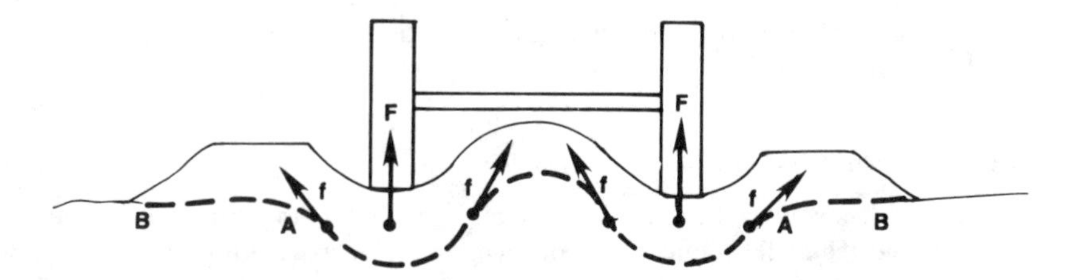

FIG. 3—*Load support provided under wheels by the "tensioned membrane effect." The Forces* F *are the resultants of the vertical components of Forces* f. *The geotextile deformation is evaluated by assuming that the subgrade deforms at constant volume, that is, the volume heaving is equal to the volume settling. Lateral anchorage is provided over the lengths* AB.

quently, prevents excessive deformations under repeated loads [*9*]. As a result, stresses on the subgrade soil can be as high as its bearing capacity (that is, plastic equilibrium, also called plastic limit). In contrast, in the case of unpaved roads without any reinforcement, stresses on the subgrade should be kept smaller than the subgrade soil elastic limit to prevent the development of excessive deformations under repeated loads. (The elastic limit is the value of the applied stress that initiates the development of a plastic zone in the subgrade soil.) If the subgrade soil is assumed to be a saturated clay ($\phi = 0$), its elastic and plastic limits are πc and $(\pi + 2)c$, respectively. Therefore by using a reinforcement with an appropriate modulus, the stresses on the subgrade can be increased by the ratio $(\pi + 2)/\pi$ [*10*]. The increase in allowable stresses on the subgrade has been confirmed by tests [*9*].

The tensioned membrane effect can be evaluated, as indicated in Fig. 3, by using the classical soil mechanics approach of a constant volume deformation for the saturated clay subgrade. A theoretical relationship has thus been established between the rut depth and the geotextile's elongation, leading to a relationship between the rut depth, the aggregate layer thickness, and the geosynthetic's tensile modulus [*10*].

Combining the analyses of confinement effect and tensioned membrane effect just outlined, charts have been established to determine the aggregate layer thickness reduction made possible by using geotextile or geogrid reinforcement [*5,6,10*].

Again, it appears that the design method for the geosynthetic application results from the adaptation of a classical soil mechanics method—bearing capacity derived from plasticity theory—using a classical soil mechanics approach: the constant-volume deformation of a saturated clay when it is subjected to rapid loading.

Designing with Geosynthetics Tomorrow Afternoon and Later

Having reviewed designs that geotechnical engineers can do as early as tomorrow morning, let us try to predict what can be done a bit later, whether tomorrow afternoon or many afternoons thereafter. Predictions can be made from information on products being developed and research being carried out. However, the following discussion cannot be complete because of proprietary restrictions and elementary discretion regarding research being carried out and concepts being evaluated. Evolution of design concepts and design methods in the future may be linked to developments occurring in the areas of available materials, construction methods, evaluation of geosynthetics properties, and research.

Materials

Range of Materials—In classical geotechnical engineering, only natural materials such as soils and rocks are used. Types and properties of natural materials are limited, while the human imagination is unlimited and can thereby produce a wide variety of products which in and of themselves may lead to new design concepts. These products include not only geotextiles but also all sorts of geosynthetics such as grids, mats, nets, meshes, etc. When discussing the future, the term "geosynthetics" should be expanded to a broader term such as "geoproducts" to include manmade products such as steel wires, cables, meshes, etc. as well as products from biotechnology. Consider for a moment the possible combinations of these products—natural, synthetic, and biologically altered; the possibilities are endless.

Geosynthetics—The growing geosynthetics market will attract more and more manufacturers. These manufacturers will come not only from the traditional textile industry like most manufacturers today, but also from the overall plastics industry. They will bring new ideas and create new products. Some of the production lines will be sufficiently versatile to allow designers to obtain custommade geosynthetics to fit special design needs. For example, synthetic drainage materials are already gaining wide acceptance. However, to provide large drainage capacity, these products must contain very large voids and are therefore bulky and difficult to transport. Expansive geosynthetics (either made of a permeable foam, or similar in principle, if not in use, to honeycomb products already on the market) that take their final shape and occupy their final volume only when expanded on the site will be developed to minimize transportation costs. The use of waste products (such as reclaimed tires) for soil reinforcement will grow, requiring the development of special testing and design methods. The development of the use of fibers, threads [*11*], and minigrids [*12*] directly combined with soil has already led to a new design concept, microreinforcement, in addition to the concept of macroreinforcement already in use. (See a comprehensive discussion in Ref *3*.)

Microgeosynthetics—Use of these small elements such as filaments, threads, and minigrids will not be limited exclusively to reinforcement. One may envision using them for filtration: there are cases where a thick filter is needed, but it is difficult to obtain close contact between a thick geotextile filter and the soil to be filtered (which is required to prevent uncontrolled movement of soil particles adjacent to the filter, which might lead to clogging). One solution to this problem would be to make the filter on site to the desired thickness by spraying many fibers or filaments onto the soil, thus assuring good contact. A further refinement of this technique could be obtained through utilization of sophisticated equipment to vary either concentration of fibers in the sprayed mixture or thickness of the mixture in designated areas. Another refinement would be to fulfill an old dream of the geotechnical engineer: the cohesive material with high permeability. This would be achieved by mixing into the soil a large number of high porosity inclusions with high mechanical interlocking with the soil. Such materials would be ideal for the construction of filters in earth dams in seismic areas, where liquefaction of granular filters during earthquakes is a major problem.

Biogeoproducts—Biotechnology will bring new tools to the geotechnical engineer of the future in the form of geobiological products with controlled growth (in size and orientation) to provide in situ soil with anisotropic reinforcement or preferential permeability and/or controlled degradation to meet time-dependent needs. Areas of applications most likely for such products would be stabilization of steep slopes, erosion control, landslide repair, and revetment of swales and driveways.

Construction Methods

New construction methods will open up new possibilities for conceptual design, especially for soil reinforcement applications.

For macroreinforcement, pretensioning (Fig. 4) will be used more and more systematically, with specialized equipment to stretch the geosynthetic in the field. Today's geotechnical engineer can utilize current construction techniques to pretension geosynthetics to a certain extent. Simply having an understanding of geosynthetic behavior enables the design engineer to specify

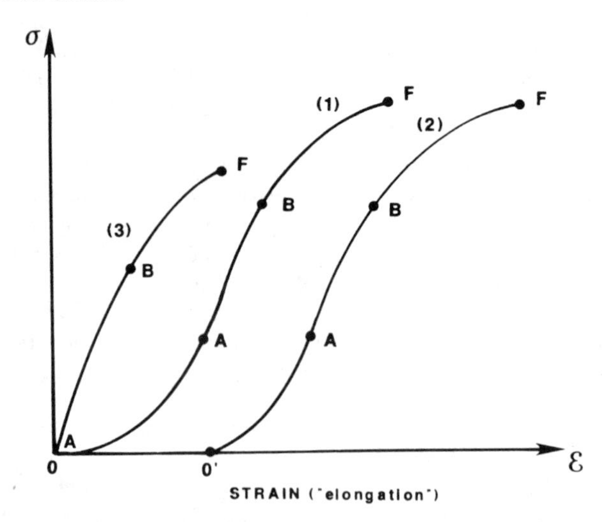

FIG. 4—*Pretensioning Concept. Both stress-strain curves are related to the same geosynthetic, initially taut (1) or wrinkled (2). Portion 00′ of Curve 2 corresponds to the elimination of the wrinkles. Pretensioning consists of stretching the geosynthetic in the field to Point A. Consequently, the geosynthetic works in the* AB *portion of the curve where its tensile modulus is high. If the same geosynthetic had been prestressed during its manufacture, and if it had been placed without wrinkles, its stress-strain curve would be Curve 3.*

installation sequences which will make the geosynthetic work more efficiently (Fig. 5). The key factor to consider in such attempts is the fact that geosynthetics have strain-dependent properties.

The developments in microreinforcement will probably have a great impact on construction methods, especially development of such items as the previously mentioned computerized equipment controlling density and orientation of microreinforcement for placement of fibers or minigrids in the ground. As a result of this technology, it will become possible to construct the "ideal least expensive structure" by placing microreinforcement in the appropriate direction with respect to the expected strain field. The corresponding method of design will have to be developed.

Many applications can be envisioned where macroreinforcement and microreinforcement will be combined. Macroreinforcement will be preferred in applications where the analysis of the

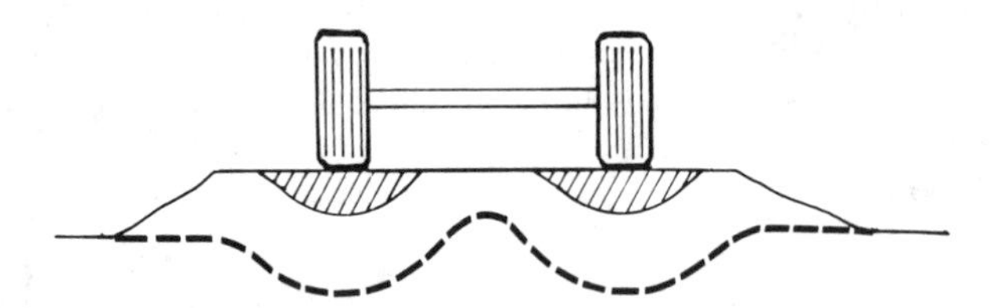

FIG. 5—*Pretensioning of a geosynthetic in unpaved road. First, an aggregate thickness smaller than designed is used, and ruts develop after a few vehicle passages. The ruts are then backfilled with aggregate. The three beneficial effects are: (1) the stretched geosynthetic has a higher modulus; (2) the tensioned membrane effect is increased since the force in the geosynthetic is increased (see Fig. 3); and (3) some aggregate is saved since aggregate design thickness is provided only under the wheels. This technique is effective only if the traffic is perfectly channelized.*

stress and strain field in the unreinforced structure shows that reinforcement is needed in one or two well-defined directions. Microreinforcement can be used in all cases but is expected to find its broadest use in applications where reinforcement is needed in all directions because tensile stresses and strains may appear in several directions or in directions that vary with time.

Use and Evaluation of Geosynthetics Properties

Computerized data banks will be used to store and retrieve properties of existing geosynthetics and, by comparing properties to selection criteria, will automatically select geosynthetics suitable for a project. The efficiency of such systems will depend on the development of international standards, but many factors impede progress in this area. For example:

1. Standards organizations are notoriously slow to develop national standards, so it will be even more difficult to establish accepted international standards.
2. Slowness of standards organizations may be even more of a problem in a fast-growing discipline. New products appear and are used—or misused—before relevant standard test methods are available. As a result, the designer's choice is too often between testing using irrelevant but available standards or not testing.
3. Efficient participation of U.S. organizations in international standardization efforts is severely hampered by the persistent use in the United States of an illogical and impractical system of units. It is of utmost importance that all influential organizations in the field of geosynthetics take a firm stance on the matter of recommending and promoting the use of the International System of Units (SI). All those who present papers, publish articles, and write brochures and reports with the exclusive use of the customary system of units should stop this practice because they do a disservice to the profession and to themselves. In our century, perpetuating a practice that restricts communication between people is a mistake from a technical, economical, and even ethical standpoint.

International standardization of symbols as well as units is critical to a smooth growth of knowledge and efficient exchange of information. Cooperation between all members of the geotechnical community is needed. One example of such cooperation is the adoption by the International Geotextile Society of a list of symbols [*13*] that is consistent with the list of symbols of the International Society for Soil Mechanics and Foundation Engineering.

Research

Developments occurring through research can be expected in two major areas: development of sophisticated analytical design methods for reinforced soil structures and development of empirical or semiempirical methods for applications for which it is not easy to develop analytical methods of design. In both cases, essential information will have to be obtained from the evaluation of the performance of carefully monitored structures.

Sophisticated Analytical Design Methods—Sophisticated analytical design methods will permit the design engineer to take into account complex parameters such as strain-dependent and time-dependent geosynthetic properties, influence of confinement stresses on geosynthetic tensile properties, soil/geosynthetics interaction, etc. The development of sophisticated analytical design methods will happen almost automatically because of these preexisting factors:

1. Such methods are particularly well adapted to soil reinforcement design, so economic incentives will cause manufacturers of geosynthetics for reinforcement to sponsor the development of such methods.
2. Many researchers are already in a position to develop such methods because they have been involved in the development of similar methods for soil mechanics applications.

3. The development of analytical methods can be done by small teams with standard equipment typically available in universities and research centers.

Such methods will require:

1. Tests to better quantify geosynthetic properties such as tensile characteristics (including creep) when confined in soil.
2. Mathematical tools such as finite-element computer programs which can model geosynthetic behavior.
3. A new approach to evaluating the safety of a structure by using a spectrum of partial factors of safety since a unique factor of safety has little meaning for a composite material such as a reinforced soil.

Results expected from this approach include:

1. Design methods based on a rational analysis for reinforced soil structures such as vertical walls, steep slopes, embankments on weak foundations, etc.
2. Methods to optimize orientation and density of microreinforcing elements (such as fibers or minigrids) to design the least expensive structure and direct the automated equipment used to place the reinforcing elements.

Development of Empirical and Semiempirical Methods—The development of analytical methods for certain applications is difficult. This is typically the case for traffic supporting structures (for example, roads, railway track structures) because fatigue of materials, natural as well as synthetic, under the effect of repeated loads is not well understood.

Highway pavements are designed today using empirical methods that result from extensive full-scale testing conducted over the past several decades. It is necessary that similar full-scale testing be undertaken to establish design methods for geosynthetic-reinforced paved and unpaved roads.

A first step towards that goal has been the development of an empirical method for the design of unpaved roads using available data from 15 years of experience [*14*]. The next step should be to carry out extensive full-scale testing for unpaved as well as paved roads. Governmental agencies, geosynthetics manufacturers, research centers, and professional and trade organizations are encouraged to get together to sponsor and conduct such research.

Empirical or semiempirical methods of design usually requires the use of a classification of materials. Classification of materials requires an accepted set of index properties and, therefore, accepted testing procedures to evaluate index properties. Unfortunately, at the present time, there is a great deal of confusion regarding geosynthetic index properties because of too many discrepancies between test procedures used by manufacturers, laboratories, research centers, etc. This confusion concerning index tests for geosynthetics must be eliminated as soon as possible through cooperation between all parties involved. Again, international cooperation is sorely needed. A first step could be the adoption by all manufacturers of a standard format to present the properties of their products.

Conclusion

As shown in the first part of this paper, using three design methods as examples, geosynthetic design methods clearly relate to the classical soil mechanics design of the same structures without geosynthetics. Considering the close bridge between classical soil mechanics and designing with geosynthetics, geotechnical engineers should be encouraged to design more and more using geosynthetics.

It can be predicted that a comprehensive and consistent body of knowledge for designing with geosynthetics will be developed in the next decade. As a result, an increasing number of large structures will be built incorporating geosynthetics. Also, use of geosynthetics will be safer than it is today because design engineers will be better informed and less likely to design exclusively on the basis of information provided by manufacturers, as many now do. An ideal situation will be reached when design engineers have several design methods at their disposal for each type of application, so they can apply their judgement.

Design engineers should be kept informed of new products, new design concepts, and new design methods. The International Society for Soil Mechanics and Foundation Engineering has recognized the need for continuing education of geotechnical engineers on geotextiles and has formed a Technical Committee on Geotextiles.

Increased cooperation between geotechnical engineers and polymer specialists (including textile specialists) is encouraged. Geotechnical engineers can help develop new products and new designs by expressing design needs and by "dreaming" of ideal solutions. Polymer and textile specialists know the possibilities and limitations of the plastics and textile industry and can therefore help in the orientation of new designs.

A geosynthetic designer's "tomorrow" may already exist somewhere else. Staying informed and exchanging ideas are the best ways to make tomorrow happen today.

Acknowledgments

The author is indebted to J. F. Beech, R. Bonaparte, J. E. Fluet, Jr., and C. Pearce for many valuable comments in the preparation of this paper.

References

[*1*] Giroud, J. P., Arman, A., and Bell, J. R., "Geotextiles in Geotechnical Engineering, Research and Practice," report of the International Society of Soil Mechanics and Foundation Engineering, Technical Committee on Geotextiles, *Geotextiles and Geomembranes,* Vol. 2, No. 3, July 1985, pp. 179–242.

[*2*] Koerner, R. M. and Bove, J. A., "Lateral Drainage Designs Using Geotextiles and Geocomposites," this publication.

[*3*] Bonaparte, R., Holtz, R. D., and Giroud, J. P., "Soil Reinforcement Design Using Geotextiles and Geogrids," this publication.

[*4*] Haliburton, T. A. and Wood, P. D., "Evaluation of the U.S. Army Corps of Engineer Gradient Ratio Test for Geotextile Performance," *Proceedings,* Second International Conference on Geotextiles, Vol. 1, Las Vegas, NV, 1982, pp. 97–101.

[*5*] Giroud, J. P., *Geotextiles and Geomembranes—Definitions, Properties and Design,* International Fabrics Association International, St. Paul, MN, 1984.

[*6*] Christopher, B. R. and Holtz, R. D., *Geotextile Engineering Manual,* Federal Highway Administration, Washington, DC, 1985.

[*7*] Heerten, G. and Wittman, L., "Filtration Properties of Geotextiles and Mineral Filters Related to River and Canal Bank Protection," *Geotextiles and Geomembranes,* Vol. 2, 1985, pp. 47–63.

[*8*] Bonaparte, R., Williams, N., and Giroud, J. P., "Innovative Leachate Collection Systems for Hazardous Waste Containment Facilities," *Proceedings of the Geotechnical Fabrics Conference,* Cincinnati, OH, June 1985, International Fabrics Association International, St. Paul, MN, pp. 9–34.

[*9*] Bender, D. A. and Barrenberg, E. J., "Design and Behavior of Soil-Fabric Aggregate Systems," *Proceedings, Transportation Research Board,* Washington, DC, Jan. 1978.

[*10*] Giroud, J. P. and Noiray, L., "Geotextile—Reinforced Unpaved Roads Design," *Journal of the Geotechnical Division,* Vol. 107, No. GT9, Proc. paper 16489, American Society of Civil Engineers, New York, Sept. 1981, pp. 1233–1254.

[*11*] Leflaive, E., "The Reinforcement of Granular Materials with Continuous Fibers," *Proceedings of the Second International Conference on Geotextiles,* Vol. 3, Las Vegas, NV, Aug. 1982, pp. 721–726.

[*12*] Mercer, F. B., Andrawes, K. Z., McGown, A., and Hytiris, N., "A New Method of Soil Stabilization," *Proceedings of the Symposium on Polymer Grid Reinforcement,* Vol. 8.1, London, March 1984.

[*13*] Giroud, J. P., "Symbols for Geotechnical Engineering, Geotextiles and Geomembranes," list of symbols adopted by the International Geotextile Society, *Geotextiles and Geomembranes,* Vols. 2, 3, July 1985, pp. 243-262.
[*14*] "Recommandations pour l'emploi des geotextiles dans les voies de circulation provisoire, les voies a faible traffic et les couches de forme," published by Comite Francais des Geotextiles et Geomembranes, Boulogne sur Seine, France, Feb. 1981.

Committee D-35 Position Paper

L. David Suits,[1] *Robert G. Carroll, Jr.,*[2] *and Barry R. Christopher*[3]

ASTM Geotextile Committee Testing Update[4]

REFERENCE: Suits, L. D., Carroll, R. G., Jr., and Christopher, B. R., **"ASTM Geotextile Committee Testing Update,"** *Geotextile Testing and the Design Engineer, ASTM STP 952,* J. E. Fluet, Jr., Ed., American Society for Testing and Materials, Philadelphia, 1987, pp. 161–175.

ABSTRACT: The use of geotextiles is a relatively newly accepted concept in the field of geotechnical engineering. As a result there is a lack of standard procedures or test methods for evaluating the engineering properties or characteristics of geotextiles. ASTM Committee D-35 on Geotextiles, Geomembranes and Related Products has evolved from two former subcommittees working jointly under two different main committees. This took place in the interest of accelerating the development of much needed test standards. The paper reviews the activities of Committee D-35 in the development of these standards. It also presents the authors' recommendations as to procedures to be followed until such time as these methods become accepted through ASTM. The basic procedures for the three standards that have been accepted and published as of the summer of 1985 are also presented.

KEY WORDS: geotextiles, standards, tensile strength, creep

ASTM Committee D-35 on Geotextiles, Geomembranes and Related Products was originally formed in 1977 as Subcommittee D13.61 of Committee D-13 on Textiles by the textile industry. The objective was to provide both manufacturers and users with testing standards for measuring physical and mechanical properties of textiles used in civil engineering. Many of the tests under development at that time by the subcommittee involved soil fabric interaction principles of geotechnical engineering. As a result, in 1980 the subcommittee (D13.61) elected to become a joint committee under ASTM Committee D-13 on Textiles and the Committee D-18 on Soil and Rock, thereby promoting more involvement from the geotechnical community. In the interest of accelerating the development of geotextile standards, the evolution of the joint subcommittee led in 1984 to the formation of the present main Committee D-35 on Geotextiles, Geomembranes and Related Products.

The purpose of this position update is

1. To review the status of current Committee D-35 activities.
2. To discuss state of the art and geotextile testing methods currently in use.

[1] Soils Engineering Laboratory Supervisor, New York State Department of Transportation, Soil Mechanics Bureau, Building 7, 1220 Washington Ave., Albany, NY 12232. Chairman of ASTM Committee D35.03 on Permeability and Filtration.

[2] Manager, Pavement Reinforcement Products, The Tensar Corp., 1210 Citizens Parkway, Morrow, GA 20260. Chairman of ASTM Committee D35.01 on Mechanical Properties.

[3] Principle engineer, STS Consultants, 111 Pfingsten Rd., Northbrook, IL 60062. Chairman of ASTM Committee D-35 on Geotextiles, Geomembranes and Related Products.

[4] ASTM Committee D-35 on Geotextiles, Geomembranes and Related Products, ASTM, 1916 Race St., Philadelphia, PA 19103. This article originally appeared in the December 1985 issue of *Geotechnical Testing Journal,* published by ASTM.

3. To recommend standard procedural points with those test methods that are already widely used in the geotextile community.

Specific step-by-step descriptions of the methods referred to in this paper are forthcoming in an ASTM special technical publication (STP).

Currently the committee has over 20 methods under development for standardization. These standards fall under three categories:

1. Mechanical properties.
2. Endurance properties.
3. Permeability and filtration properties.

The committee has subcommittees devoted to each of these test development categories, along with a fourth subcommittee to define terminology related to geotextile standards.

Committee D-35 has approved three standards to date. The first standards to be produced by Committee D-35 were done so in the spring of 1984. These are: (1) ASTM Test Method for Deterioration of Geotextiles from Exposure to Ultraviolet Light and Water (Xenon Arc Type Apparatus) (D 4355) and ASTM Practice for Sampling of Geotextiles for Testing (D 4354). A third standard was produced in the spring of 1985 for the Water Permeability of Geotextiles, ASTM Permittivity Method (D 4491). Several other standards are near completion; however, much more development and review effort are required to provide the broad range of standards necessary to the industry. It is a paramount task for this committee to evolve test standards that are pertinent to the civil engineers' needs, technically accurate, and that can pass the scrutiny of hundreds of reviewers from many different professional fields (that is, civil engineers, textile engineers, chemists, and so forth).

Despite the technical communication problems between the disciplines, the committee has made progress toward this goal. Even so, it is realized that the geotextile community can not wait for final adoption of ASTM standards. Therefore the authors believe it imperative to establish a position on the current state of practice in geotextile testing.

Geotextiles are being used widely throughout the world with fabric specifiers already adopting a variety of standards to meet their immediate needs. Those methods currently used come from a variety of sources. Some are existing ASTM or American Association of Textile Chemists and Colorists (AATCC) standards, while others have been developed by both manufacturers and users. Many of the textile test methods were developed to evaluate other types of products and, therefore, do not always address specific concerns of geotextile users. Despite their origin, these standards have been used for over a decade as a means of characterizing geotextile properties for product comparison, and in some cases, for design purposes. Despite the apparent standard procedures for these textile standards, there is usually considerable room for interpretation, or options, in the procedural points in many of them. As a result, two or more labs might perform the same test using different, but allowable, test parameters, and report results that are not comparable. Committee D-35 believes that it is imperative to revise these commonly used test methods to provide a consistent methodology for all who use them with geotextiles.

The following presents the status and position of the various test methods presently under development by each subcommittee of D-35.

Subcommittee D35.01 on Mechanical Properties

The Mechanical Properties Subcommittee of D-35 is devoted to test development for measuring physical and mechanical properties of geotextiles, for example, mass per unit area, strength, and so forth. The subcommittee's current priorities are grab tensile strength, wide width strip tensile strength, trapezoid tear strength, and seam strength. The previously mentioned adopted standard on sampling falls under the jurisdiction of this subcommittee.

The standardization of sampling techniques are critical to the statistical validity of any test evaluation, and, therefore, Subcommittee D35.01 has established a standard for sampling as its first method. Trapezoid tear and grab tensile strength are current ASTM standards that require modification to assure uniform procedural points for all geotextile test laboratories. In addition to modifying existing ASTM standards, the section is developing a wide strip tension strength test that will reduce the effects of "fabric neck down" (Poisson's effect), which biases test results in favor of some fabric types.

The wide width strip tension test provides a more reliable assessment of fabric tensile strength than the grab method. Wide strip tensile strength values are reported in units of Newtons per metre (pound per inch) and are considered useful to the design engineer in calculating a fabric reinforcement value.

The major problem with already existing ASTM strength test standards is that these methods offer options in their procedural points, that is, test equipment, clamp geometry, sample size, strain rate, and reporting format. The authors believe that it is mandatory to revise these current ASTM standards by limiting each method to specific procedural points, without options, to assure a unified testing standard between all geotextile labs. The authors, therefore, make the following recommendations regarding these currently used standards.

Grab Tensile Strength/ASTM D 1682

The current ASTM standard Test Methods for Breaking Load and Elongation of Textile Fabrics [D 1682-64 (1975)] lists four tension strength tests: grab, ravel strip, cut strip, and modified grab. Of these methods, the grab method has been selected as the standard of practice by the geotextile community to provide a qualitative assessment of geotextile strength. Within ASTM D 1682 there are three options for test machines, two options for strain rate, and two options for clamp size. For performing the grab strength test on geotextiles, the authors recommend the method be conducted according to the procedures defined in ASTM D 1682, Section 16, using the following fixed procedural points:

1. Tensile testing machine . . . use only constant rate of extension (CRE) testing machine or one capable of applying the load in the CRE mode, for example, Instron.
2. Strain rate . . . load to break at a strain rate of 300 mm/min (12 in./min); no option for "constant" time to break.
3. Jaw faces . . . each clamp shall have one jaw face measuring 25.4 mm (1 in.) perpendicular to the direction of application of the load and 50 mm (2 in.) parallel to the direction of application of the load. The opposing jaw face of each clamp shall be at least as large as its mate. Each jaw face shall be in line both with respect to its mate on the same clamp and on the corresponding jaw on the other clamp. No modifications to jaw size are allowed.
4. Sample size . . . sample size shall be 100 mm wide by 200 mm gage length (4 by 8 in.).

Wide Strip Tension Test D 4595-86

The wide width strip tension test is one developed by the subcommittee specifically for geotextiles [ASTM Test Method for Tensile Properties of Geotextiles by the Wide Width Strip Method (D 4595-86)]. Conventional strip and grab tension tests call for narrow strips with aspect ratios (that is, ratio of clamped width to gage length) less than one. These tests have been deemed inappropriate to produce stress strain behavior of geotextiles. The wide width strip tension test with a specimen aspect ratio of 2:1 appears to provide a more realistic measure of tensile strength and related stress strain behavior for geotextiles caused by the larger specimen size, uniformity of load distribution across the entire width of the specimen, and strain rates more in line with geotechnical applications. In this test a relatively wide specimen is gripped across its entire width in the clamps of a CRE type tension testing machine operated at a prescribed rate of extension, applying a longitudinal load to the specimen until the specimen ruptures. Tensile strength, elongation, initial and secant modulus, and breaking toughness of the specimen can be calculated from machine scales, dials, recording charts, or an interfaced computer. Following are procedural specifications recommended.

1. *Tensile Testing Machine*—A constant rate of extention (CRE) type of testing machine described in ASTM Specification for Tensile Testing Machines for Textiles (D 76-77) shall be used. When using the CRE type tensile tester, the recorder should have adequate pen response to properly record the load-elongation curve as specified in ASTM Standard D 76.
2. *Specimen Size*—200 mm (8 in.) wide by 100 mm (4 in.) gage length.
3. *Clamps*—Those which are sufficiently wide to grip the entire width of the sample and with appropriate clamping power to prevent slipping or crushing (damage). Wedge type clamps have been found to provide the most acceptable clamping action. Two basic clamp designs are shown in Figs. 1 through 4. These designs have been used in the laboratory and have provided acceptable tensile strengths. These clamps may be modified to provide greater ease and speed of clamping providing the basic wedge design is maintained.
4. *Size of Jaw Faces*—Each clamp shall have jaw faces measuring wider than the width of the specimen, 200 mm (8 in.), and provide a minimum of 25 mm (1 in.) clamping length on the specimen in the direction of the load.
5. *Strain Rate*—A strain rate of 10%/min (0.4 in./min, 10 mm/min).

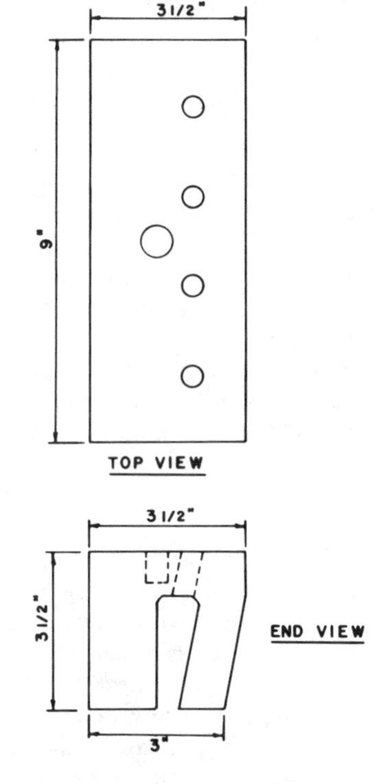

FIG. 1—*Wide width test clamps and Christopher clamp: top and end view.*

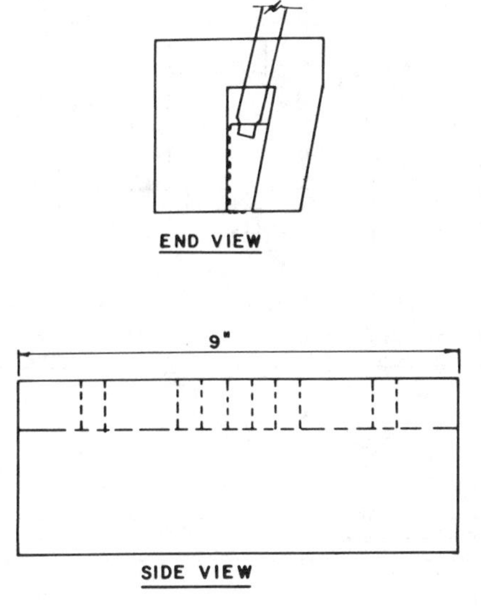

FIG. 2—*Wide width test clamps and Christopher clamp: end and side view.*

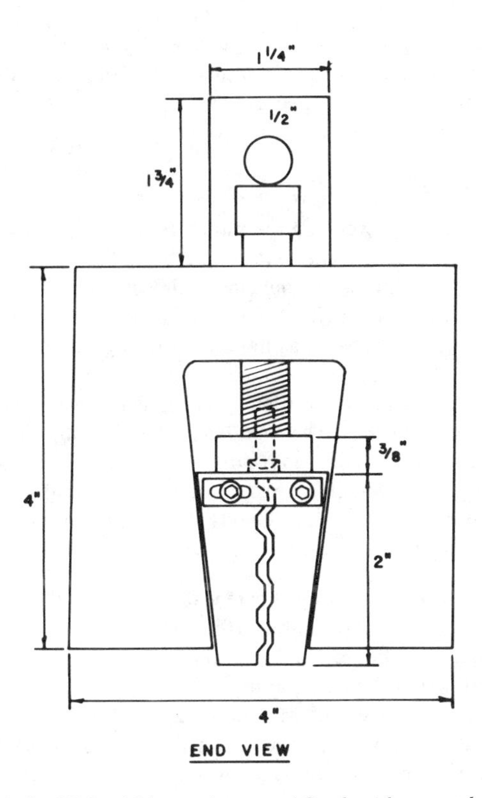

FIG. 3—*Wide width test clamps and Sanders clamp: end view.*

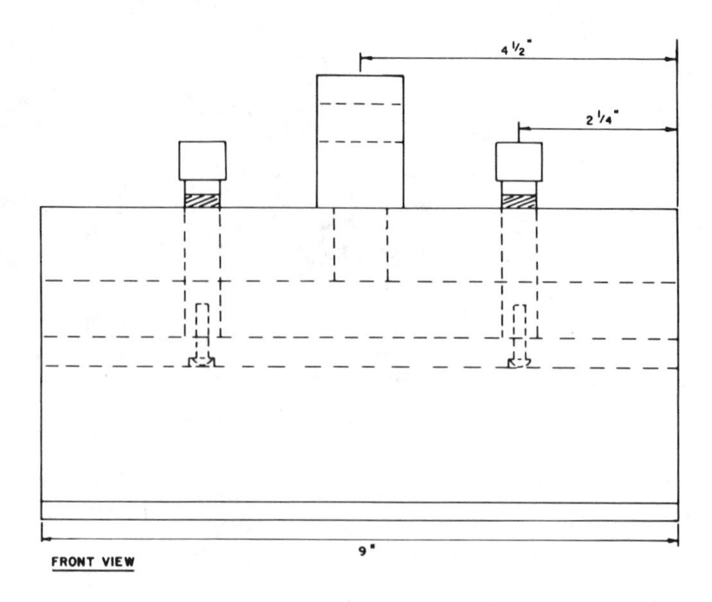

FIG. 4—*Wide width test clamps and Sanders clamp: front view.*

The subcommittee is in the process of round-robin evaluation of the wide width strip tension testing concept. The round-robin will be evaluating interlab reproducibility, clamping effects, sample dimensions, strain rates, and aspect ratios.

Puncture Test/ASTM D 3787

The Corps of Engineers first developed the puncture test for geotextiles in the early 1970s. The original Corps method specified the following: ASTM Method of Testing Coated Fabrics (D 751-79) tension testing machine with ring clamp and replacing the steel ball with an 8-mm (5/16-in.)-diameter solid steel cylinder (flat tip) with a flat centered within the ring clamp. Since its origin, the U.S. Corps of Engineers has modified the original puncture test to use a hemispherical tip probe rather than a flat tip probe to penetrate the geotextile. The geotextile community has experienced a great deal of difficulty with this test when using a hemispherical tip probe on high-strength woven fabrics. The rounded probe often slips between yarns rather than rupturing yarns, thus leading to erroneous results.

The authors recommend that the puncture test be run in general accordance to ASTM Test Method for Bursting Strength of Knitted Goods: Constant-Rate-of Transverse (CRT), Ball Burst Test (D 3787-80*a*) with the following modifications:

1. Testing machine . . . a constant rate of extension (CRE) or other test machines capable of applying a tensile or compressive load in the CRE mode.
2. Puncture attachment . . . the steel ball replaced with a 8-mm (5/16-in.) diameter solid steel flat tip cylinder centered within the ring clamp.
3. Strain rate . . . a strain rate of 300 mm/min (12 in./min).
4. In addition to the type of clamps described in ASTM D 3787, compression rings have been used that allow for testing in compression-type testing machines. This method appears acceptable.

Trapezoid Tear/ASTM D 4533

The authors recommend the use of the trapezoid tear method as defined in ASTM Test Method for Trapezoid Tearing Strength of Geotextiles (D 4533-85) for evaluating a fabric's resistance to propagating a tear with the following fixed procedural points and modifications:

1. Tension test machine . . . use only constant rate of extension (CRE) test apparatus or one capable of applying a load in the CRE mode.
2. Strain rate . . . a strain rate of 300 mm/min (12 in./min).
3. Specimen size . . . as per the test method, the specimens shall be rectangular in shape cut to 75 by 200 mm (3 by 8 in.). Cut the specimens to be used for the measurement of the tearing strength in the machine direction with the longer dimension parallel to the machine direction and the specimens to be used for the measurement in the tearing strength in the cross machine direction with a longer dimension parallel to the cross matching direction.
4. Calculating tear load . . . record the maximum force in Newtons (pounds) as illustrated in Fig. 5. Note: this is a modification to ASTM D 1117.
5. Clamp location . . . clamp along the nonparallel sides of the trapezoid so that the end edges of the clamps are in line with the 25-mm (1-in.) side of the trapezoid.

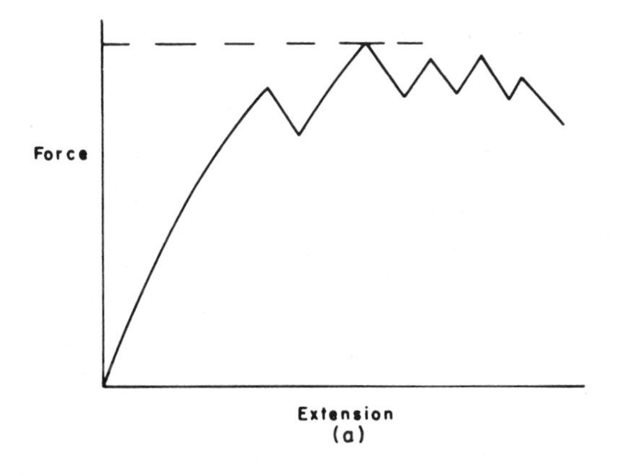

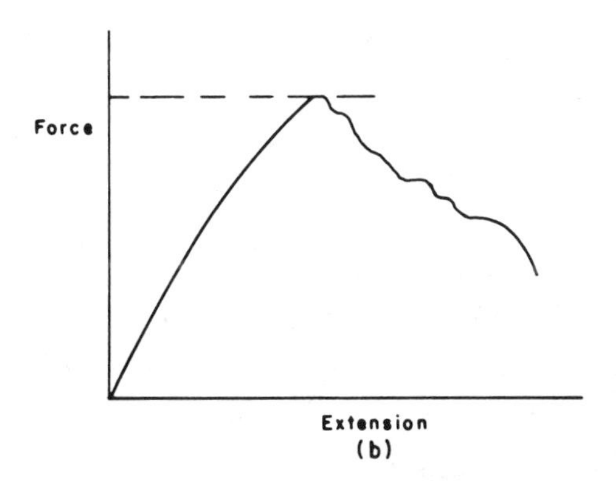

FIG. 5—*Typical tearing force—Extension curves for individual test specimens:* (a) *fabric exhibiting several maxima and* (b) *fabric exhibiting single maximum.*

Diaphragm Bursting Strength Test Method/ASTM D 3786

The Mullen burst (diaphragm burst) test is widely used in the geotextile industry as a measure of fabric resistance to rupture from load by hydraulic diaphragm. The test listed as ASTM Test Method for Hydraulic Bursting Strength of Knitted Goods and Nonwoven Fabrics: Diaphragm Bursting Strength Tester Method (D 3786-80*a*). The authors recommend that the diaphragm burst test be run according to procedural points defined in ASTM D 3786 with the following modifications regarding the bursting diaphragm and report on results:

1. *Diaphragms*—A diaphragm of molded synthetic rubber 1.80 ± 0.05 mm (0.070 ± 0.002 in.) in thickness with reinforced center, clamped between the lower clamping plate and the rest of the apparatus so that before the diaphragm is stretched by pressure underneath it the center of its upper surface is below the planes of the clamping surface. The pressure required to raise the free surface of the diaphragm plane shall be 30 ± 6 kPa (4.3 ± 0.8 psi). This pressure shall be checked at least once a month. To test, a bridge gage may be used, the test being carried out with the clamping ring removed. The diaphragm should be inspected frequently for permanent distortion and renewed if necessary.
2. *Report*—Report bursting strength as the maximum bursting pressure registered on the pressure gage. Do not subtract the tare pressure associated with the diaphragm.

Subcommittee D35.02 on Endurance Properties

The endurance property subcommittee was established to develop and standardize test methods that evaluate ability of a geotextile to withstand the adverse effects of stress and environment encountered during installation and in service life. This subcommittee has identified eight categories of endurance standards for development. The subcommittee has one standard already approved in the area of the effects of ultraviolet light on geotextiles. The Test Method for Deterioration of Geotextiles from Exposure to Ultraviolet Light and Water (Xenon-Arc Type Apparatus) (D 4355) was accepted and approved in the spring of 1984.

The following is a list of the endurance properties subcommittee's test methods presently under development:

1. Abrasion resistance.
2. Creep behavior.
3. Chemical resistance.
4. On site protection and handling.
5. Temperature stability.
6. Biological stability.
7. Fatigue resistance.

Abrasion Resistance

The subcommittee has had great difficulty in agreeing on the best conceptual approach to evaluate abrasion resistance in geotextiles. The difficulty lies in simulating field conditions. Four test methods have been investigated for possible abrasion techniques:

1. Sand sieve vibrator to simulate large aggregate ballast abrasion of geotextiles.
2. Modified Stoll flex abrader (a reciprocating sliding block-type abrader) to simulate small aggregate abrasion of geotextiles.
3. Los Angeles drum abrader to simulate large aggregate/ballast abrasion of geotextiles.
4. Rotary platform—double head (Taber abrader) to measure the abrasion resistance of geotextiles to loaded rotating wheels.

The most widely used abrasion test in most current practice is the rotary platform double head method. This test was originally defined in ASTM D 1175, but has been revised and redesignated by ASTM as Test Method for Abrasion Resistance of Textile Fabrics (Rotary Platform, Double-Head Method) (D 3884-80). This abrasion test method requires a specimen be abraded using a rotary rubbing action under controlled conditions of pressure and abrasive action. The specimen is mounted on the platform and turns on a vertical axis against the sliding rotation of two abrading wheels. One abrading wheel rubs the specimen outward toward the periphery and the other inward towards the center. The resulting abrasion marks form a circular pattern over an area approximately 30 cm^2. Resistance to abrasion is evaluated by tension tests on abraded and controlled samples to determine strength retention.

This method does not directly simulate the type of abrasive action that a geotextile experiences in use but the test does provide a useful index to fabric abrasion resistance that has been used for nearly a decade by geotextile specifiers. Even so, the subcommittee has found difficulty in obtaining reproducible results and, as such, has rejected this procedure.

Since a replacement test has not been established, the authors recognize that some specifications call for Taber abrasion. When this is the case the authors recommend that the following fixed parameters be used:

1. A rubber-based, CS-17, calibrase abrasive wheel, manufactured by Tabor Instrument Co., or equivalent.
2. A load on each abrasive wheel of 10 N (equivalent to a 1000 g mass).
3. Remove the center pin from the abrasion specimen holder.
4. To determine residual breaking strength use the 50-mm (2-in.) raveled or cut strip tension test as specified in ASTM Test Methods for Breaking Load and Elongation of Textile Fabrics [D 1682-64 (1975)], using a constant rate of extension type tension machine, and strain rate of 300 mm/min (12 in./min) except that the distance between the clamps shall be 25.4 mm (1 in.). Horizontally place the path of abrasion on the abraded specimen midway between the clamps.
5. Unless otherwise specified, strength retention should be measured after 500 and 1000 cycles.

Biological and Chemical Stability

The biological and chemical stability of geotextiles is a common concern for many users/specifiers, despite the relative inertness of the synthetic materials used to manufacture geotextiles to the ambient conditions of most geotextile applications. The subcommittee is working on standards to address these concerns since there are no suitable standards available. The textile industry does, however, have a basic knowledge in polymer chemistry that allows them to evaluate the suitability of their synthetic materials in most biological and chemical environments.

In most geotextile applications, biological and chemical stability is not a real problem. Synthetic fabrics, that is, polypropylene and polyester are nonbiodegradeable and highly resistant to a broad range of acid and alkaline conditions. Natural soil conditions rarely contain a caustic element detrimental to geotextile stability. There may be special cases, however, where a geotextile might be used in a potentially caustic environment causing strength losses or changes in hydraulic properties (for example, hazardous waste, disposal sites, and petroleum tank yards). In such cases, it is advisable for the user/specifier to consult with a qualified professional (for example, polymer chemist) regarding fabric suitability for the specific site conditions.

The problem with a caustic environment is to determine the actual elements or combination of elements that may come in contact with a geotextile. If the caustic elements cannot be clearly defined, it is advisable to run lab or field trials that expose a fabric to the potentially caustic condition in a manner that simulates actual use conditions. Fabric samples can then be extracted at specific intervals, for example, 1 week, 1 month, and so forth, to test for deterioration of strength or other pertinent properties. Such trials are laborious and require extended test periods to achieve meaningful results but are more reliable than accelerated tests since correlations have not been made to laboratory tests.

Creep Behavior

Of concern to the geotechnical engineer in designing the use of geotextiles is the long-term strength and creep behavior of geotextiles. The subcommittee is presently developing a method for determining the creep characteristics of geotextiles. However, it is felt that in order to develop an appropriate creep standard there must be a standard method for determining the tensile strength characteristics of the geotextile. Therefore work on this standard is being held up until such time that a standard method for determining the tensile strength characteristics of a geotextile is adopted. The main point in question is the specimen size to be tested. Also, much work has been done in studying creep behavior in the confined and unconfined conditions. At this point in time the authors make no recommendations as to a creep standard method of testing.

The authors suggest a 50.8-mm (2-in.) wide specimen be tested for a qualitative assessment of creep limit and the wide strip method be used as a quantitative method to analyze creep rate.

On-Site Protection and Handling

One of the chief concerns and problems in the use of geotextiles is the manner in which they are handled and stored on-site while awaiting installation. Therefore, the subcommittee is developing a practice that can be followed for on-site protection and handling. Examples of this are:

1. Proper labeling and identification.
2. Protection against ultraviolet degradation while awaiting installation.
3. Method of storage.

Ultraviolet Degradation Resistance

As previously mentioned, an ultraviolet degradation test method has been adopted and is presently in use. The test procedure requires geotextile specimens to be exposed for 0, 150, 300, and 500 h of ultraviolet (UV) exposure in a Xenon arc device. The exposure consists of 120-min cycles as follows:

1. 102 min of light only, at 65 ± 5°C (150 ± 10°F) black panel temperature and 30% relative humidity.
2. 18 min of water spray and light.

Following each exposure time, the respective weathered specimen is subjected to a 50.8-mm (2-in.) cut or raveled strip tension test as defined in ASTM Test Method for Breaking Load and Elongation of Textile Fabrics (D 1682) using a CRE tensile machine and a strain rate of 300 mm/min (12 in./min). A plot of the strengths versus hours of UV exposure will provide an indication of the geotextiles tendency to deteriorate.

Note that geotextiles are manufactured using a variety of polymers, and each polymer varies in its sensitivity to UV radiation. UV radiation from the sun varies with duration of exposure, angle of inclination of the sun, topography, and geography. The Xenon arc test cannot simulate all these variables. So it is not likely that the Xenon arc test results will be directly proportional to a sunlight exposure test but will provide an indication as to degradation tendency.

Subcommittee D35.03 on Permeability and Filtration

The permeability and filtration subcommittee was established to develop methods to characterize drainage and filtration related properties or behavior of geotextiles.

Listed as follows are the methods or standards that the subcommittee is presently developing:

1. Permittivity.
2. Apparent opening size.
3. Gradient ratio (soil fabric permeability).
4. Transmissivity under load and under no load.
5. Permittivity under load.
6. Thickness.

Permittivity and Hydraulic Conductivity

The objective of the permeability tests is to define a value of fabric permeability. The coefficient of permeability k, using Darcy's equation, has been the most commonly reported value. The coefficient of permeability of fabric is used in comparison with the coefficient of permeability for a protected soil to assure compatibility, that is, k (fabric) $>$ k (soil). The measurement of the coefficient of permeability is dependent on the fabric thickness and therefore is not an appropriate factor for comparing relative flow capacities of fabrics. It is possible for geotextiles of different thicknesses to have the same Darcy coefficient of permeability, while having large differences in their flow capacity for water.

In dealing with the permeability of the geotextile alone, there have been many different test methods used by both manufacturers and users in determining the permeability of the geotextile. These include constant head and falling head tests. There are many variables that have to be considered when performing these tests. In the constant head test, head differentials ranged from 50 mm to 1 m (2 in. to 3 ft) or greater. The falling head tests may range from a 600-mm (2-ft) change to a 50-mm (2-in.) change in head. There are those methods that use deaired water, and some correct for temperature, bringing back to 20°C, while there are methods that do neither. Testing has also been done comparing multiple thickness of the geotextile to single thickness, but it was found that there was no direct proportional relationship between the results for single and multiple thicknesses of geotextile.

Each of the just-cited procedural variations has been found to produce significant variations in test results.

As a result of a round-robin test program in which eleven laboratories took part, decisions on the variables previously mentioned have been made and have been incorporated into the accepted standard. The results of the round-robin showed the following:

1. As long as the device did not control the test any design in apparatus would be appropriate. Recommended minimum specimen diameter is 50 mm (2 in.).
2. In the constant head test, heads of 50 mm (2 in.) provided good reproducibility.
3. In the falling head test, as long as the head loss was comparable to the 50-mm (2-in.) head in the constant head test, the two methods provide comparable results.
4. Test water deaired to a maximum of 6 ppm of desolved oxygen allowed reproducibility of tests.
5. To provide ability to compare results between laboratories, the end result should be corrected back to the 20°C (68°F).

As previously mentioned this standard has been adopted. The test yields a value that is not dependent on the fabric thickness, that is, permittivity. Permittivity of a geotextile is defined as the volumetric flow rate of water per unit cross-sectional area per unit head under laminar flow conditions, in the normal direction through a geotextile. The units are seconds to the minus one ($cm^3/s/cm^2/cm = s^{-1}$).

Apparent Opening Size

Soil retention is a predominant function of geotextiles in drainage and filtration applications. Fabric pore size is the key parameter that controls a fabrics ability to retain the soils. Calhoun [*1*], with the U.S. Army/Corp of Engineers Waterways Experiment Station, developed a test for equivalent opening size (EOS) to characterize the soil particle retention ability of the various fabrics. The test involved sieving rounded sand particles of a specified size through the fabric to determine that fraction of particle sizes for which 5% or less, by weight, passed through the fabric. The EOS was defined as the "retained on" size of that fraction expressed as a U.S. standard sieve number (for example, No. 70 [210 μm]). Assuming that fabrics and screen mesh have comparable retention ability, the EOS was a rational means of correlating fabric pore structure to an equivalent screen mesh size.

It is imperative to note that EOS values do not accurately define fabric pore sizes or pore structure. The EOS test only provides a method for determining the relative size of the largest straight through openings in a fabric [*2*]. Note that two fabrics may have similar EOS values but dramatically different pore structures and porosities, for example, woven versus nonwoven fabrics. Despite these deficiencies EOS testing does provide a useful index value for predicting a fabric's soil retention ability. The authors recommend the EOS method with revised procedural points. The method has been renamed "apparent opening size" (AOS).

A round-robin program was performed to evaluate reproducibility of the AOS test. The program addressed the following factors in order to generate more reproducible results:

1. Surface films.
2. Static buildup on glass beads.
3. Order of bead range sieving (fine to coarse or coarse to fine).
4. Shaking time.
5. Variability between specimens.

In some instances, the manufacturing process leaves a surface coating on the geotextile, which may act to clog some of the openings. This leads to erroneous test results in that beads that would normally pass through the geotextile cannot. However, for most geotextiles that have this coating, it has been found that soaking the fabric in water will remove this coating, allowing the appropriate size beads to pass.

The buildup of static electricity causes glass beads to cling to the geotextile rather than pass through, also leading to erroneous results. Static eliminators added to sieve frame walls will reduce or eliminate the buildup of static electricity (Fig. 6). Static elimination has been accomplished with some success by using commercially available static eliminators such as spray, sheets used in laundry dryers, or static master devices attached to the sieve frame.

Round-robin testing revealed that when the AOS test begins with coarse glass beads, certain fabric pores that would pass finer beads might become clogged. The round-robin program also noted no significant difference in test results for varying sieve times from 5 to 20 min.

Taking all these factors into consideration, the proposed method for AOS will require the following:

1. A 1-h presoak of the specimen in distilled water followed by drying at 30 ± 5°C (86 ± 10°F) until a constant weight is reached.

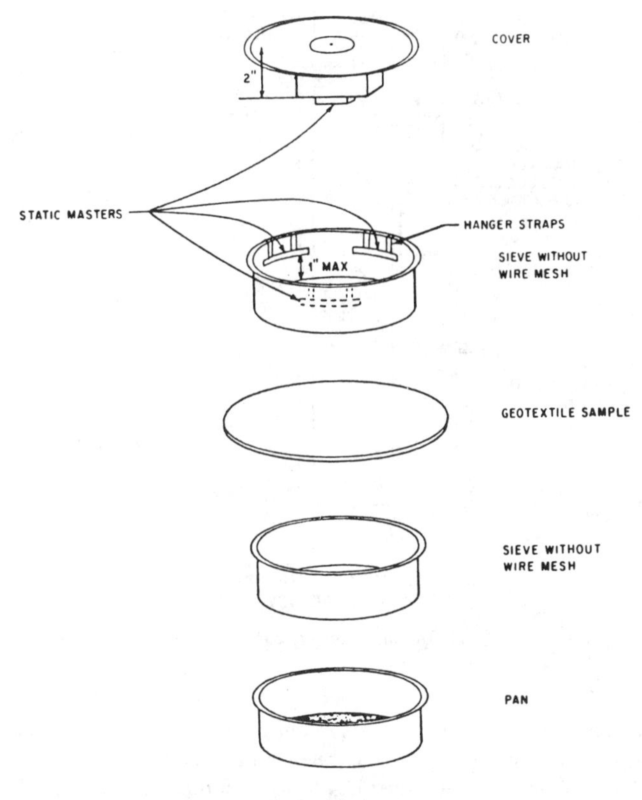

FIG. 6—*Sieving test.*

2. Drying of the glass beads at the same temperature.
3. Static elimination techniques are to be used in the sieving frame to decrease or eliminate the buildup of static electricity on the beads.
4. The sieving process will start with fine beads first, progressing to the coarse beads.
5. A 10-min shaking time is proposed.
6. All size glass beads are shaken through one specimen.

Gradient Ratio (Soil-Geotextile System Performance)

Clogging potential is a major concern of geotextile users. To truly evaluate a geotextile clogging potential, a soil-fabric permeability test must be developed to evaluate soil-fabric interaction under specified soil and hydraulic conditions.

In the early 1960s and later modified by the U.S. Army Corp of Engineers in the early 1970s, a soil-fabric permeability system test was developed that reportedly measured a fabrics clogging potential in terms of a "clogging ratio." The clogging ratio test, as defined by the Corp, does not provide sufficient procedural details for operating under standard conditions and test parameters. The subcommittee feels that this testing concept has merit and is developing a standard method for review and round-robin evaluation, that is, gradient ratio.

The procedure calls for placing the geotextile and 76.2 to 101.6 mm (3 to 4 in.) of a select soil in a 101.6-mm (4 in.) diameter permeameter apparatus (Fig. 7). Several piezometers or manometers are placed at various heights above the geotextile in the soil, as well as below the

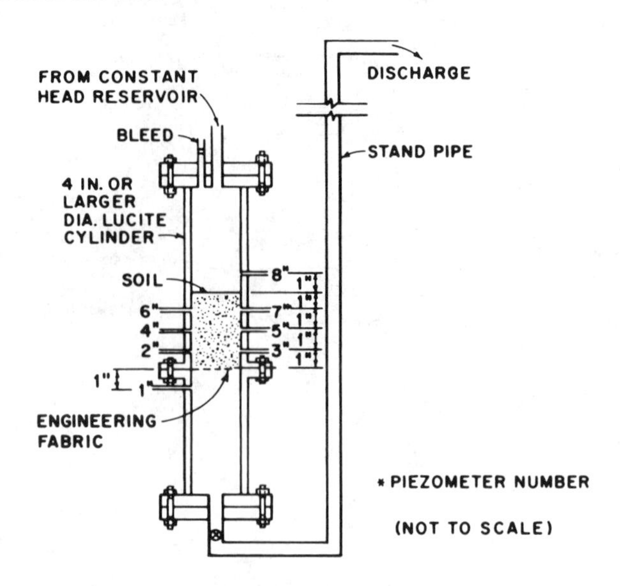

FIG. 7—*Detail of constant head permeameter test device used for gradient ratio testing.*

geotextile. A constant head of water is applied to cause flow through the system, and head losses are measured at each piezometer location at regular time intervals up to 24 h. Once the flow stabilizes, the flow is allowed to continue to 24 h. At 24 h the final readings are taken and the gradient ratio determined for the final readings.

Gradient ratio is determined by dividing the hydraulic gradient of the lower 25 mm (1 in.) of soil plus geotextile by the hydraulic gradient through the adjacent 55 mm (2 in.) of soil, between 25 and 5 mm (1 and 3 in.) above the geotextile. The U.S. Corp of Engineers specification calls for a gradient ratio equal to or less than 3 as the allowable limit for acceptable performance.

As previously indicated, the subcommittee is evaluating this test through a round-robin program. The program includes analysis of procedure, apparatus, reproducibility of results, and interpretation of results.

In-Plane Transmissivity (Underload)

Another concern in the performance of geotextiles is the ability of water to flow through the plane of the geotextile. The subcommittee is presently preparing draft standards that will describe techniques for determining this ability of a geotextile. Until such time that these methods have been drafted and balloted by the subcommittee, no particular recommendation is being made as to their use.

Permittivity Underload

The subcommittee is presently preparing a draft standard for determining the permittivity of geotextiles under loading. This proposed standard will essentially test a geotextile using the previously described permittivity procedure with the addition of a normal load placed on the geotextile. It is felt that this type of test will be useful as a design mechanism in determining the performance of geotextiles in the field.

Since this proposed procedure has not been balloted by the subcommittee as of this date, no recommendation is being made as to a standard method.

Thickness

While it appears that the characteristic of thickness of a geotextile is more of a mechanical property than it is a permeability and filtration property, it is necessary to know what geotextile thickness is being tested in the in-plane transmissivity and the permittivity under load test methods. Therefore the thickness standard has been assigned to the Permeability and Filtration Subcommittee for development. The nominal thickness of a geotextile can be measured using ASTM Method for Measuring Thickness of Textile Materials [D 1777-64 (1975)]; however, this procedure does not specify the normal pressure to be used. A normal pressure of 2 kPa (7.3 psi) has been recommended as a standard for the nominal thickness of geotextiles and must be specified with the ASTM procedure. A draft standard is currently being prepared to determine the compressed geotextile thickness.

Summary

There are numerous tests in addition to those discussed here being used in the geotextile industry to characterize fabrics. This paper addresses only those that the authors believe to be most controversial and in need of immediate clarification. The majority of test methods referenced in this paper are index tests, that is, the results generated are not design values, but index values that allow for the relative comparisons of geotextiles. These tests are widely used throughout the geotextile industry for specification purposes. This paper should in no way be construed as an endorsement by the authors to the relevance or applicability of the methods discussed. The paper is merely an attempt to provide unity in testing methods while the committee's own standards are being established.

Acknowledgments

The authors wish to thank the Executive Subcommittee of ASTM D-35 for the review and comments received in the preparation of this paper.

References

[1] Calhoun, C. C., "Development of Design Criteria and Acceptance Specification for Plastic Filter Cloth," Army Corps of Engineers, Waterways Experiment Station, Vicksburg, MS, June 1972.

[2] Carroll, R. G., Jr., "Geotextile Filter Criteria," prepared for 1983 Symposium on Geotextiles, sponsored by the Transportation Research Board Task Force on Engineering Fabrics—A2T57, Washington, DC, Jan. 1983.

Subject Index

D

E

F

R

S

Author Index